Introduction to Modern Power Electronics

Introduction to Modern Power Electronics

Edited by
Giani Smith

Larsen & Keller
www.larsen-keller.com

Introduction to Modern Power Electronics
Edited by Giani Smith
ISBN: 978-1-63549-231-6 (Hardback)

Larsen & Keller

Published by Larsen and Keller Education,
5 Penn Plaza,
19th Floor,
New York, NY 10001, USA

Cataloging-in-Publication Data

Introduction to modern power electronics / edited by Giani Smith.
    p. cm.
Includes bibliographical references and index.
ISBN 978-1-63549-231-6
1. Power electronics. 2. Electronics. 3. Electric power. I. Smith, Giani.
TK7881.15 .I58 2017
621.317--dc23

The publisher's policy is to use permanent paper from mills that operate a sustainable forestry policy. Furthermore, the publisher ensures that the text paper and cover boards used have met acceptable environmental accreditation standards.

Printed and bound in the United States of America.

For more information regarding Larsen and Keller Education and its products, please visit the publisher's website www.larsen-keller.com

Table of Contents

Preface

This book elucidates the concepts and innovative models around prospective developments with respect to power electronics. It describes in detail the uses and applications of this subject in the present-day scenario. Power electronics refers to the practice of controlling and converting electrical power by using solid state electronics. It is integrally connected to electrical engineering and commercial management of electronic systems. This text attempts to understand the multiple topics that fall under the discipline of power electronics and how such concepts have practical applications. It is a compilation of chapters that discuss the most vital concepts in this field. While understanding the long-term perspectives of the topics, the book makes an effort in highlighting their impact as a modern tool for the growth of the discipline. It will serve as a valuable source of reference for those interested in this field.

To facilitate a deeper understanding of the contents of this book a short introduction of every chapter is written below:

Chapter 1- Power electronics are devices that control and convert electricity like rectifiers, inverters, AC-to-AC converters and similar appliances. This chapter provides a comprehensive overview of power electronics and studies devices that are essential to power electronics like switched-mode power supply and commutation cell. The chapter on power electronics offers an insightful focus, keeping in mind the complex subject matter.

Chapter 2- The generation of electric power has been the focal point of civilization and the history of electricity generation can be traced back to as long ago as 6th century BC. This chapter briefs the reader about mechanisms and methods of electric energy generation like wind turbine, hydroelectricity, induction generator, solar cell, tidal stream generator and windbelt. The section on wind turbine also talks about the modification of wind lens. The content is tailored to provide elaborate information on the devices and their operations and applications.

Chapter 3- Semiconductor devices utilize semiconductor materials and their conductivity can be regulated or manipulated by doping or the application of an electromagnetic field, light or even heat. This chapter explores semiconductor devices like power semiconductor device, diode, zener diode, tunnel diode, thyristor and transistor. The content focuses on the operation, construction and uses of these devices. This chapter elucidates the crucial theories and applications of semiconductor devices.

Chapter 4- Inverters are appliances that convert power from DC to AC. This chapter studies the power inverter, AC/AC converter (its types like cycloconverter and sparse matrix converter, solar inverter, DC-to-DC converter and rectifier. The content explores the circuit description, size, applications and input and output of each. Power electronics is best understood in confluence with the major topics listed in the following chapter.

Chapter 5- Switched-mode power supply is a method of regulating yield voltage or current by switching elements of the circuit thereby increasing efficiency. The chapter explores fluorescent lamps, battery charger and switched-mode power supply and provides the principles of operation and applications of each. The aspects elucidated in this chapter are of vital importance, and provide a better understanding of power electronics.

I owe the completion of this book to the never-ending support of my family, who supported me throughout the project.

Editor

Introduction to Power Electronics

Power electronics are devices that control and convert electricity like rectifiers, inverters, AC-to-AC converters and similar appliances. This chapter provides a comprehensive overview of power electronics and studies devices that are essential to power electronics like switched-mode power supply and commutation cell. The chapter on power electronics offers an insightful focus, keeping in mind the complex subject matter.

Power Electronics

Power electronics is the application of solid-state electronics to the control and conversion of electric power. It also refers to a subject of research in electronic and electrical engineering which deals with the design, conttrol, computation and integration of nonlinear, time-varying energy-processing electronic systems with fast dynamics.

An HVDC thyristor valve tower 16.8 m tall in a hall at Baltic Cable AB in Sweden

The first high power electronic devices were mercury-arc valves. In modern systems the conversion is performed with semiconductor switching devices such as diodes, thyristors and transistors, pioneered by R. D. Middlebrook and others beginning in the 1950s. In contrast to electronic systems concerned with transmission and processing of signals and data, in power electronics substantial amounts of electrical energy are processed. An AC/DC converter (rectifier) is the most typical power electronics device found in many consumer electronic devices, e.g. television sets, personal computers, battery chargers, etc. The power range is typically from tens of watts to several hundred watts. In industry a common application is the variable speed drive (VSD) that is used to control an induction motor. The power range of VSDs start from a few hundred watts and end at tens of megawatts.

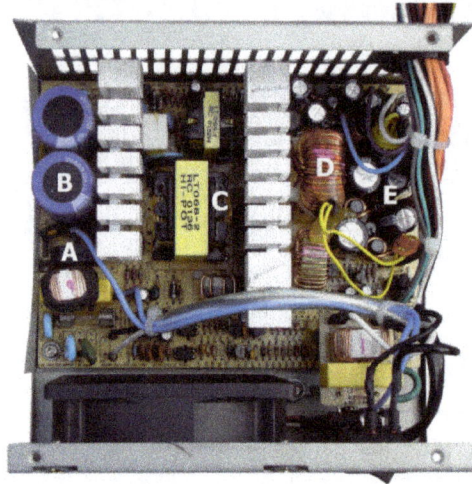

A PCs power supply is an example of a piece of power electronics, whether inside or outside of the cabinet

The power conversion systems can be classified according to the type of the input and output power

- AC to DC (rectifier)
- DC to AC (inverter)
- DC to DC (DC-to-DC converter)
- AC to AC (AC-to-AC converter)

History

Power electronics started with the development of the mercury arc rectifier. Invented by Peter Cooper Hewitt in 1902, it was used to convert alternating current (AC) into direct current (DC). From the 1920s on, research continued on applying thyratrons and grid-controlled mercury arc valves to power transmission. Uno Lamm developed a mercury valve with grading electrodes making them suitable for high voltage direct current power transmission. In 1933 selenium rectifiers were invented.

In 1947 the bipolar point-contact transistor was invented by Walter H. Brattain and John Bardeen under the direction of William Shockley at Bell Labs. In 1948 Shockley's invention of the bipolar junction transistor (BJT) improved the stability and performance of transistors, and reduced costs. By the 1950s, higher power semiconductor diodes became available and started replacing vacuum tubes. In 1956 the silicon controlled rectifier (SCR) was introduced by General Electric, greatly increasing the range of power electronics applications.

By the 1960s the improved switching speed of bipolar junction transistors had allowed for high frequency DC/DC converters. In 1976 power MOSFETs became commercially available. In 1982 the Insulated Gate Bipolar Transistor (IGBT) was introduced.

Devices

The capabilities and economy of power electronics system are determined by the active devices that are available. Their characteristics and limitations are a key element in the design of power electronics systems. Formerly, the mercury arc valve, the high-vacuum and gas-filled diode thermionic rectifiers, and triggered devices such as the thyratron and ignitron were widely used in power electronics. As the ratings of solid-state devices improved in both voltage and current-handling capacity, vacuum devices have been nearly entirely replaced by solid-state devices.

Power electronic devices may be used as switches, or as amplifiers. An ideal switch is either open or closed and so dissipates no power; it withstands an applied voltage and passes no current, or passes any amount of current with no voltage drop. Semiconductor devices used as switches can approximate this ideal property and so most power electronic applications rely on switching devices on and off, which makes systems very efficient as very little power is wasted in the switch. By contrast, in the case of the amplifier, the current through the device varies continuously according to a controlled input. The voltage and current at the device terminals follow a load line, and the power dissipation inside the device is large compared with the power delivered to the load.

Several attributes dictate how devices are used. Devices such as diodes conduct when a forward voltage is applied and have no external control of the start of conduction. Power devices such as silicon controlled rectifiers and thyristors (as well as the mercury valve and thyratron) allow control of the start of conduction, but rely on periodic reversal of current flow to turn them off. Devices such as gate turn-off thyristors, BJT and MOSFET transistors provide full switching control and can be turned on or off without regard to the current flow through them. Transistor devices also allow proportional amplification, but this is rarely used for systems rated more than a few hundred watts. The control input characteristics of a device also greatly affect design; sometimes the control input is at a very high voltage with respect to ground and must be driven by an isolated source.

As efficiency is at a premium in a power electronic converter, the losses that a power electronic device generates should be as low as possible.

Devices vary in switching speed. Some diodes and thyristors are suited for relatively slow speed and are useful for power frequency switching and control; certain thyristors are useful at a few kilohertz. Devices such as MOSFETS and BJTs can switch at tens of kilohertz up to a few megahertz in power applications, but with decreasing power levels. Vacuum tube devices dominate high power (hundreds of kilowatts) at very high frequency (hundreds or thousands of megahertz) applications. Faster switching devices minimize energy lost in the transitions from on to off and back, but may create problems with radiated electromagnetic interference. Gate drive (or equivalent) circuits must be designed to supply sufficient drive current to achieve the full switching speed possible with a device. A device without sufficient drive to switch rapidly may be destroyed by excess heating.

Practical devices have non-zero voltage drop and dissipate power when on, and take some time to pass through an active region until they reach the "on" or "off" state. These losses are a significant part of the total lost power in a converter.

Power handling and dissipation of devices is also a critical factor in design. Power electronic devices may have to dissipate tens or hundreds of watts of waste heat, even switching as efficiently as possible between conducting and non-conducting states. In the switching mode, the power controlled is much larger than the power dissipated in the switch. The forward voltage drop in the conducting state translates into heat that must be dissipated. High power semiconductors require specialized heat sinks or active cooling systems to manage their junction temperature; exotic semiconductors such as silicon carbide have an advantage over straight silicon in this respect, and germanium, once the main-stay of solid-state electronics is now little used due to its unfavorable high temperature properties.

Semiconductor devices exist with ratings up to a few kilovolts in a single device. Where very high voltage must be controlled, multiple devices must be used in series, with networks to equalize voltage across all devices. Again, switching speed is a critical factor since the slowest-switching device will have to withstand a disproportionate share of the overall voltage. Mercury valves were once available with ratings to 100 kV in a single unit, simplifying their application in HVDC systems.

The current rating of a semiconductor device is limited by the heat generated within the dies and the heat developed in the resistance of the interconnecting leads. Semiconductor devices must be designed so that current is evenly distributed within the device across its internal junctions (or channels); once a "hot spot" develops, breakdown effects can rapidly destroy the device. Certain SCRs are available with current ratings to 3000 amperes in a single unit.

Solid-State Devices

Device	Description	Ratings
Diode	Uni-polar, uncontrolled, switching device used in applications such as rectification and circuit directional current control. Reverse voltage blocking device, commonly modeled as a switch in series with a voltage source, usually 0.7 VDC. The model can be enhanced to include a junction resistance, in order to accurately predict the diode voltage drop across the diode with respect to current flow.	Up to 3000 amperes and 5000 volts in a single silicon device. High voltage requires multiple series silicon devices.
Silicon-controlled rectifier (SCR)	This semi-controlled device turns on when a gate pulse is present and the anode is positive compared to the cathode. When a gate pulse is present, the device operates like a standard diode. When the anode is negative compared to the cathode, the device turns off and blocks positive or negative voltages present. The gate voltage does not allow the device to turn off.	Up to 3000 amperes, 5000 volts in a single silicon device.
Thyristor	The thyristor is a family of three-terminal devices that include SCRs, GTOs, and MCT. For most of the devices, a gate pulse turns the device on. The device turns off when the anode voltage falls below a value (relative to the cathode) determined by the device characteristics. When off, it is considered a reverse voltage blocking device.	

Gate turn-off thyristor (GTO)	The gate turn-off thyristor, unlike an SCR, can be turned on and off with a gate pulse. One issue with the device is that turn off gate voltages are usually larger and require more current than turn on levels. This turn off voltage is a negative voltage from gate to source, usually it only needs to be present for a short time, but the magnitude s on the order of 1/3 of the anode current. A snubber circuit is required in order to provide a usable switching curve for this device. Without the snubber circuit, the GTO cannot be used for turning inductive loads off. These devices, because of developments in IGCT technology are not very popular in the power electronics realm. They are considered controlled, uni-polar and bi-polar voltage blocking.	
Triac	The triac is a device that is essentially an integrated pair of phase-controlled thyristors connected in inverse-parallel on the same chip. Like an SCR, when a voltage pulse is present on the gate terminal, the device turns on. The main difference between an SCR and a Triac is that both the positive and negative cycle can be turned on independently of each other, using a positive or negative gate pulse. Similar to an SCR, once the device is turned on, the device cannot be turned off. This device is considered bi-polar and reverse voltage blocking.	
Bipolar junction transistor (BJT)	The BJT cannot be used at high power; they are slower and have more resistive losses when compared to MOSFET type devices. To carry high current, BJTs must have relatively large base currents, thus these devices have high power losses when compared to MOSFET devices. BJTs along with MOSFETs, are also considered unipolar and do not block reverse voltage very well, unless installed in pairs with protection diodes. Generally, BJTs are not utilized in power electronics switching circuits because of the I^2R losses associated with on resistance and base current requirements. BJTs have lower current gains in high power packages, thus requiring them to be set up in Darlington configurations in order to handle the currents required by power electronic circuits. Because of these multiple transistor configurations, switching times are in the hundreds of nanoseconds to microseconds. Devices have voltage ratings which max out around 1500 V and fairly high current ratings. They can also be paralleled in order to increase power handling, but must be limited to around 5 devices for current sharing.	
Power MOSFET	The main benefit of the power MOSFET is that the base current for BJT is large compared to almost zero for MOSFET gate current. Since the MOSFET is a depletion channel device, voltage, not current, is necessary to create a conduction path from drain to source. The gate does not contribute to either drain or source current. Turn on gate current is essentially zero with the only power dissipated at the gate coming during switching. Losses in MOSFETs are largely attributed to on-resistance. The calculations show a direct correlation to drain source on-resistance and the device blocking voltage rating, BV_{dss}. Switching times range from tens of nanoseconds to a few hundred microseconds, depending on the device. MOSFET drain source resistances increase as more current flows through the device. As frequencies increase the losses increase as well, making BJTs more attractive. Power MOSFETs can be paralleled in order to increase switching current and therefore overall switching power. Nominal voltages for MOSFET switching devices range from a few volts to a little over 1000 V, with currents up to about 100 A or so. Newer devices may have higher operational characteristics. MOSFET devices are not bi-directional, nor are they reverse voltage blocking. ||	

Insulated-gate bipolar transistor (IGBT)	These devices have the best characteristics of MOSFETs and BJTs. Like MOSFET devices, the insulated gate bipolar transistor has a high gate impedance, thus low gate current requirements. Like BJTs, this device has low on state voltage drop, thus low power loss across the switch in operating mode. Similar to the GTO, the IGBT can be used to block both positive and negative voltages. Operating currents are fairly high, in excess of 1500 A and switching voltage up to 3000 V. The IGBT has reduced input capacitance compared to MOSFET devices which improves the Miller feedback effect during high dv/dt turn on and turn off.	
MOS-controlled thyristor (MCT)	The MOS-controlled thyristor is thyristor like and can be triggered on or off by a pulse to the MOSFET gate. Since the input is MOS technology, there is very little current flow, allowing for very low power control signals. The device is constructed with two MOSFET inputs and a pair of BJT output stages. Input MOSFETs are configured to allow turn on control during positive and negative half cycles. The output BJTs are configured to allow for bidirectional control and low voltage reverse blocking. Some benefits to the MCT are fast switching frequencies, fairly high voltage and medium current ratings (around 100 A or so).	
Integrated gate-commutated thyristor (IGCT)	Similar to a GTO, but without the high current requirements to turn on or off the load. The IGCT can be used for quick switching with little gate current. The devices high input impedance largely because of the MOSFET gate drivers. They have low resistance outputs that don't waste power and very fast transient times that rival that of BJTs. ABB Group company has published data sheets for these devices and provided descriptions of the inner workings. The device consists of a gate, with an optically isolated input, low on resistance BJT output transistors which lead to a low voltage drop and low power loss across the device at fairly high switching voltage and current levels. An example of this new device from ABB shows how this device improves on GTO technology for switching high voltage and high current in power electronics applications. According to ABB, the IGCT devices are capable of switching in excess of 5000 VAC and 5000 A at very high frequencies, something not possible to do efficiently with GTO devices.	

DC/AC Converters (Inverters)

DC to AC converters produce an AC output waveform from a DC source. Applications include adjustable speed drives (ASD), uninterruptable power supplies (UPS), active filters, Flexible AC transmission systems (FACTS), voltage compensators, and photovoltaic generators. Topologies for these converters can be separated into two distinct categories: voltage source inverters and current source inverters. Voltage source inverters (VSIs) are named so because the independently controlled output is a voltage waveform. Similarly, current source inverters (CSIs) are distinct in that the controlled AC output is a current waveform.

Being static power converters, the DC to AC power conversion is the result of power switching devices, which are commonly fully controllable semiconductor power switches. The output waveforms are therefore made up of discrete values, producing fast transitions rather than smooth ones. The ability to produce near sinusoidal waveforms around the fundamental frequency is dictated by the modulation technique controlling when, and for how long, the power valves are on and off. Common modulation techniques include the carrier-based technique, or Pulse-width modulation, space-vector technique, and the selective-harmonic technique.

Voltage source inverters have practical uses in both single-phase and three-phase applications. Single-phase VSIs utilize half-bridge and full-bridge configurations, and are widely used for power supplies, single-phase UPSs, and elaborate high-power topologies when used in multicell configurations. Three-phase VSIs are used in applications that require sinusoidal voltage waveforms, such as ASDs, UPSs, and some types of FACTS devices such as the STATCOM. They are also used in applications where arbitrary voltages are required as in the case of active filters and voltage compensators.

Current source inverters are used to produce an AC output current from a DC current supply. This type of inverter is practical for three-phase applications in which high-quality voltage waveforms are required.

A relatively new class of inverters, called multilevel inverters, has gained widespread interest. Normal operation of CSIs and VSIs can be classified as two-level inverters, due to the fact that power switches connect to either the positive or to the negative DC bus. If more than two voltage levels were available to the inverter output terminals, the AC output could better approximate a sine wave. It is for this reason that multilevel inverters, although more complex and costly, offer higher performance.

Each inverter type differs in the DC links used, and in whether or not they require freewheeling diodes. Either can be made to operate in square-wave or pulse-width modulation (PWM) mode, depending on its intended usage. Square-wave mode offers simplicity, while PWM can be implemented several different ways and produces higher quality waveforms.

Voltage Source Inverters (VSI) feed the output inverter section from an approximately constant-voltage source.

The desired quality of the current output waveform determines which modulation technique needs to be selected for a given application. The output of a VSI is composed of discrete values. In order to obtain a smooth current waveform, the loads need to be inductive at the select harmonic frequencies. Without some sort of inductive filtering between the source and load, a capacitive load will cause the load to receive a choppy current waveform, with large and frequent current spikes.

There are three main types of VSIs:

- Single-phase half-bridge inverter

- Single-phase full-bridge inverter

- Three-phase voltage source inverter

Single-Phase Half-Bridge Inverter

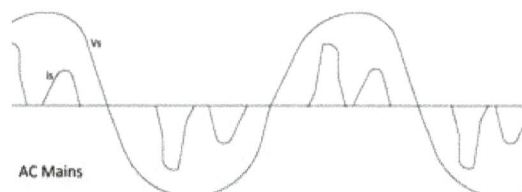

Figure 8: The AC input for an ASD.

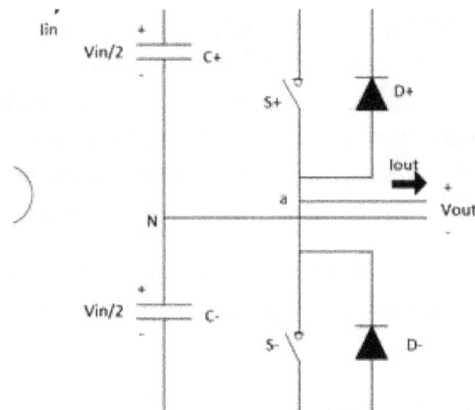

FIGURE 9: Single-Phase Half-Bridge Voltage Source Inverter

The single-phase voltage source half-bridge inverters, are meant for lower voltage applications and are commonly used in power supplies. Figure 9 shows the circuit schematic of this inverter.

Low-order current harmonics get injected back to the source voltage by the operation of the inverter. This means that two large capacitors are needed for filtering purposes in this design. As Figure 9 illustrates, only one switch can be on at time in each leg of the inverter. If both switches in a leg were on at the same time, the DC source will be shorted out.

Inverters can use several modulation techniques to control their switching schemes. The carrier-based PWM technique compares the AC output waveform, v_c, to a carrier voltage signal, v_Δ. When v_c is greater than v_Δ, S+ is on, and when v_c is less than v_Δ, S- is on. When the AC output is at frequency fc with its amplitude at v_c, and the triangular carrier signal is at frequency f_Δ with its amplitude at v_Δ, the PWM becomes a special sinusoidal case of the carrier based PWM. This case is dubbed sinusoidal pulse-width modulation (SPWM).For this, the modulation index, or amplitude-modulation ratio, is defined as $m_a = v_c/v_\Delta$.

The normalized carrier frequency, or frequency-modulation ratio, is calculated using the equation $m_f = f_\Delta/f_c$.

If the over-modulation region, ma, exceeds one, a higher fundamental AC output voltage will be observed, but at the cost of saturation. For SPWM, the harmonics of the output waveform are at well-defined frequencies and amplitudes. This simplifies the design of the filtering components needed for the low-order current harmonic injection from the operation of the inverter. The maximum output amplitude in this mode of operation is half of the source voltage. If the maximum output amplitude, m_a, exceeds 3.24, the output waveform of the inverter becomes a square wave.

As was true for PWM, both switches in a leg for square wave modulation cannot be turned on at the same time, as this would cause a short across the voltage source. The switching scheme requires that both S+ and S- be on for a half cycle of the AC output period. The fundamental AC output amplitude is equal to $v_{o1} = v_{aN} = 2v_i/\pi$.

Its harmonics have an amplitude of $v_{oh} = v_{o1}/h$.

Therefore, the AC output voltage is not controlled by the inverter, but rather by the magnitude of the DC input voltage of the inverter.

Using selective harmonic elimination (SHE) as a modulation technique allows the switching of the inverter to selectively eliminate intrinsic harmonics. The fundamental component of the AC output voltage can also be adjusted within a desirable range. Since the AC output voltage obtained from this modulation technique has odd half and odd quarter wave symmetry, even harmonics do not exist. Any undesirable odd (N-1) intrinsic harmonics from the output waveform can be eliminated.

Single-Phase Full-Bridge Inverter

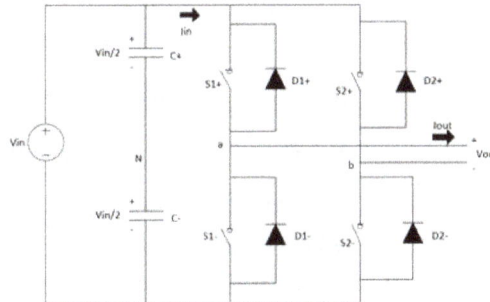

FIGURE 3: Single-Phase Voltage Source Full-Bridge Inverter

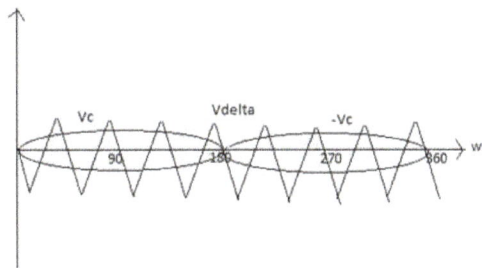

FIGURE 4: Carrier and Modulating Signals for the Bipolar Pulsewidth Modulation Technique

The full-bridge inverter is similar to the half bridge-inverter, but it has an additional leg to connect the neutral point to the load. Figure 3 shows the circuit schematic of the single-phase voltage source full-bridge inverter.

To avoid shorting out the voltage source, S1+ and S1- cannot be on at the same time, and S2+ and S2- also cannot be on at the same time. Any modulating technique used for the full-bridge configuration should have either the top or the bottom switch of each leg on at any given time. Due to the extra leg, the maximum amplitude of the output waveform is Vi, and is twice as large as the maximum achievable output amplitude for the half-bridge configuration.

States 1 and 2 from Table 2 are used to generate the AC output voltage with bipolar SPWM. The AC output voltage can take on only two values, either Vi or −Vi. To generate these same states using a half-bridge configuration, a carrier based technique can be used. S+ being on for the half-bridge corresponds to S1+ and S2- being on for the full-bridge. Similarly, S- being on for the half-bridge corresponds to S1- and S2+ being on for the full bridge. The output voltage for this modulation technique is more or less sinusoidal, with a fundamental component that has an amplitude in the linear region of less than or equal to one $v_{o1} = v_{ab1} = v_i \cdot m_a$.

Unlike the bipolar PWM technique, the unipolar approach uses states 1, 2, 3 and 4 from Table 2 to generate its AC output voltage. Therefore, the AC output voltage can take on the values Vi, 0 or −V i. To generate these states, two sinusoidal modulating signals, Vc and −Vc, are needed, as seen in Figure 4.

Vc is used to generate VaN, while −Vc is used to generate VbN. The following relationship is called unipolar carrier-based SPWM $v_{o1} = 2 \cdot v_{aN_1} = v_i \cdot m_a$.

The phase voltages VaN and VbN are identical, but 180 degrees out of phase with each other. The output voltage is equal to the difference of the two phase voltages, and do not contain any even harmonics. Therefore, if mf is taken, even the AC output voltage harmonics will appear at normalized odd frequencies, fh. These frequencies are centered on double the value of the normalized carrier frequency. This particular feature allows for smaller filtering components when trying to obtain a higher quality output waveform.

As was the case for the half-bridge SHE, the AC output voltage contains no even harmonics due to its odd half and odd quarter wave symmetry.

Three-Phase Voltage Source Inverter

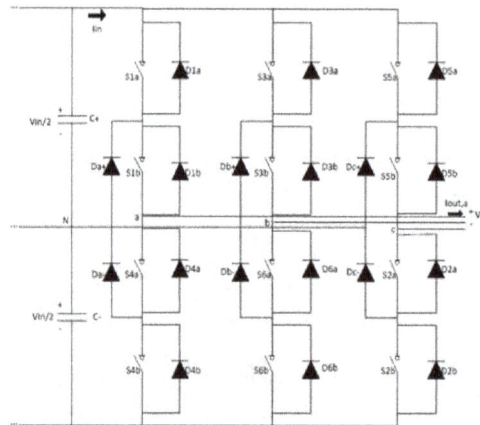

FIGURE 5: Three-Phase Voltage Source Inverter Circuit Schematic
FIGURE 6: Three-Phase Square-Wave Operation a) Switch State S1 b) Switch State S3 c) S1 Output d) S3 Output

Single-phase VSIs are used primarily for low power range applications, while three-phase VSIs cover both medium and high power range applications. Figure 5 shows the circuit schematic for a three-phase VSI.

Switches in any of the three legs of the inverter cannot be switched off simultaneously due to this resulting in the voltages being dependent on the respective line current's polarity. States 7 and 8 produce zero AC line voltages, which result in AC line currents freewheeling through either the upper or the lower components. However, the line voltages for states 1 through 6 produce an AC line voltage consisting of the discrete values of Vi, 0 or −Vi.

For three-phase SPWM, three modulating signals that are 120 degrees out of phase with one another are used in order to produce out of phase load voltages. In order to preserve the PWM features with a single carrier signal, the normalized carrier frequency, mf, needs to be a multiple of three. This keeps the magnitude of the phase voltages identical, but out of phase with each other by 120 degrees. The maximum achievable phase voltage amplitude in the linear region, ma less than or equal to one, is $v_{phase} = v_i / 2$. The maximum achievable line voltage amplitude is $V_{ab1} = v_{ab} \cdot \sqrt{3} / 2$

The only way to control the load voltage is by changing the input DC voltage.

Current Source Inverters

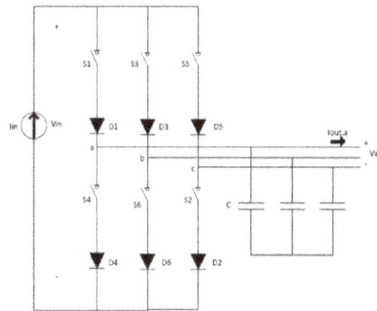

FIGURE 7: Three-Phase Current Source Inverter

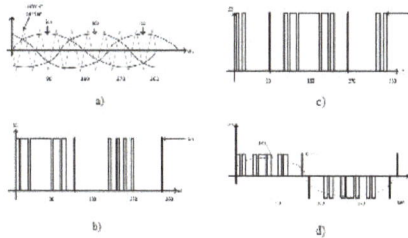

Figure 8: Synchronized-Pulse-Width-Modulation Waveforms for a Three-Phase Current Source Inverter a) Carrier and Modulating Signals b) S1 State c) S3 State d) Output Current

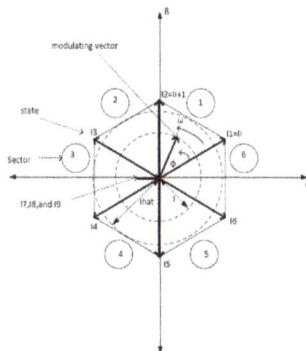

Figure 9: Space-Vector Representation in Current Source Inverters

Current source inverters convert DC current into an AC current waveform. In applications requiring sinusoidal AC waveforms, magnitude, frequency, and phase should all be controlled. CSIs have high changes in current overtime, so capacitors are commonly employed on the AC side, while inductors are commonly employed on the DC side. Due to the absence of freewheeling diodes, the power circuit is reduced in size and weight, and tends to be more reliable than VSIs. Although single-phase topologies are possible, three-phase CSIs are more practical.

In its most generalized form, a three-phase CSI employs the same conduction sequence as a six-pulse rectifier. At any time, only one common-cathode switch and one common-anode switch are on.

As a result, line currents take discrete values of $-i_i$, 0 and i_i. States are chosen such that a desired waveform is outputted and only valid states are used. This selection is based on modulating techniques, which include carrier-based PWM, selective harmonic elimination, and space-vector techniques.

Carrier-based techniques used for VSIs can also be implemented for CSIs, resulting in CSI line currents that behave in the same way as VSI line voltages. The digital circuit utilized for modulating signals contains a switching pulse generator, a shorting pulse generator, a shorting pulse distributor, and a switching and shorting pulse combiner. A gating signal is produced based on a carrier current and three modulating signals.

A shorting pulse is added to this signal when no top switches and no bottom switches are gated, causing the RMS currents to be equal in all legs. The same methods are utilized for each phase, however, switching variables are 120 degrees out of phase relative to one another, and the current pulses are shifted by a half-cycle with respect to output currents. If a triangular carrier is used with sinusoidal modulating signals, the CSI is said to be utilizing synchronized-pulse-width-modulation (SPWM). If full over-modulation is used in conjunction with SPWM the inverter is said to be in square-wave operation.

The second CSI modulation category, SHE is also similar to its VSI counterpart. Utilizing the gating signals developed for a VSI and a set of synchronizing sinusoidal current signals, results in symmetrically distributed shorting pulses and, therefore, symmetrical gating patterns. This allows any arbitrary number of harmonics to be eliminated. It also allows control of the fundamental line current through the proper selection of primary switching angles. Optimal switching patterns must have quarter-wave and half-wave symmetry, as well as symmetry about 30 degrees and 150 degrees. Switching patterns are never allowed between 60 degrees and 120 degrees. The current ripple can be further reduced with the use of larger output capacitors, or by increasing the number of switching pulses.

The third category, space-vector-based modulation, generates PWM load line currents that equal load line currents, on average. Valid switching states and time selections are made digitally based on space vector transformation. Modulating signals are represented as a complex vector using a transformation equation. For balanced three-phase sinusoidal signals, this vector becomes a fixed module, which rotates at a frequency, ω. These space vectors are then used to approximate the modulating signal. If the signal is between arbitrary vectors, the vectors are combined with the zero vectors I_7, I_8, or I_9. The following equations are used to ensure that the generated currents and the current vectors are on average equivalent.

Multilevel Inverters

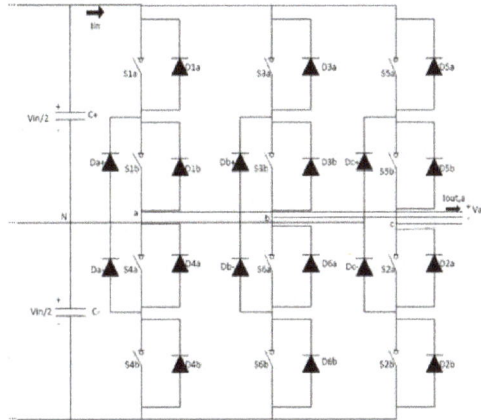

FIGURE 10: Three-Level Neutral-Clamped Inverter

A relatively new class called multilevel inverters has gained widespread interest. Normal operation of CSIs and VSIs can be classified as two-level inverters because the power switches connect to either the positive or the negative DC bus. If more than two voltage levels were available to the inverter output terminals, the AC output could better approximate a sine wave. For this reason multilevel inverters, although more complex and costly, offer higher performance. A three-level neutral-clamped inverter is shown in Figure 10.

Control methods for a three-level inverter only allow two switches of the four switches in each leg to simultaneously change conduction states. This allows smooth commutation and avoids shoot through by only selecting valid states. It may also be noted that since the DC bus voltage is shared by at least two power valves, their voltage ratings can be less than a two-level counterpart.

Carrier-based and space-vector modulation techniques are used for multilevel topologies. The methods for these techniques follow those of classic inverters, but with added complexity. Space-vector modulation offers a greater number of fixed voltage vectors to be used in approximating the modulation signal, and therefore allows more effective space vector PWM strategies to be accomplished at the cost of more elaborate algorithms. Due to added complexity and number of semiconductor devices, multilevel inverters are currently more suitable for high-power high-voltage applications. This technology reduces the harmonics hence improves overall efficiency of the scheme.

AC/AC Converters

Converting AC power to AC power allows control of the voltage, frequency, and phase of the waveform applied to a load from a supplied AC system . The two main categories that can be used to separate the types of converters are whether the frequency of the waveform is changed. AC/AC converter that don't allow the user to modify the frequencies are known as AC Voltage Controllers, or AC Regulators. AC converters that allow the user to change the frequency are simply referred to as frequency converters for AC to AC conversion. Under frequency converters there are three different types of converters that are typically used: cycloconverter, matrix converter, DC link converter (aka AC/DC/AC converter).

AC voltage controller: The purpose of an AC Voltage Controller, or AC Regulator, is to vary the RMS voltage across the load while at a constant frequency. Three control methods that are generally accepted are ON/OFF Control, Phase-Angle Control, and Pulse Width Modulation AC Chopper Control (PWM AC Chopper Control). All three of these methods can be implemented not only in single-phase circuits, but three-phase circuits as well.

- ON/OFF Control: Typically used for heating loads or speed control of motors, this control method involves turning the switch on for n integral cycles and turning the switch off for m integral cycles. Because turning the switches on and off causes undesirable harmonics to be created, the switches are turned on and off during zero-voltage and zero-current conditions (zero-crossing), effectively reducing the distortion.

- Phase-Angle Control: Various circuits exist to implement a phase-angle control on different waveforms, such as half-wave or full-wave voltage control. The power electronic components that are typically used are diodes, SCRs, and Triacs. With the use of these components, the user can delay the firing angle in a wave which will only cause part of the wave to be outputted.

- PWM AC Chopper Control: The other two control methods often have poor harmonics, output current quality, and input power factor. In order to improve these values PWM can be used instead of the other methods. What PWM AC Chopper does is have switches that turn on and off several times within alternate half-cycles of input voltage.

Matrix converters and cycloconverters: Cycloconverters are widely used in industry for ac to ac conversion, because they are able to be used in high-power applications. They are commutated direct frequency converters that are synchronised by a supply line. The cycloconverters output voltage waveforms have complex harmonics with the higher order harmonics being filtered by the machine inductance. Causing the machine current to have fewer harmonics, while the remaining harmonics causes losses and torque pulsations. Note that in a cycloconverter, unlike other converters, there are no inductors or capacitors, i.e. no storage devices. For this reason, the instantaneous input power and the output power are equal.

- Single-Phase to Single-Phase Cycloconverters: Single-Phase to Single-Phase Cycloconverters started drawing more interest recently because of the decrease in both size and price of the power electronics switches. The single-phase high frequency ac voltage can be either sinusoidal or trapezoidal. These might be zero voltage intervals for control purpose or zero voltage commutation.

- Three-Phase to Single-Phase Cycloconverters: There are two kinds of three-phase to single-phase cycloconverters: 3φ to 1φ half wave cycloconverters and 3φ to 1φ bridge cycloconverters. Both positive and negative converters can generate voltage at either polarity, resulting in the positive converter only supplying positive current, and the negative converter only supplying negative current.

With recent device advances, newer forms of cycloconverters are being developed, such as matrix converters. The first change that is first noticed is that matrix converters utilize bi-directional, bipolar switches. A single phase to a single phase matrix converter consists of a matrix of 9 switches connecting the three input phases to the tree output phase. Any input phase and output phase can be connected together at any time without connecting any two switches from the same phase at the same time; otherwise this will cause a short circuit of the input phases. Matrix converters are lighter, more compact and versatile than other converter solutions. As a result, they are able to achieve higher levels of integration, higher temperature operation, broad output frequency and natural bi-directional power flow suitable to regenerate energy back to the utility.

The matrix converters are subdivided into two types: direct and indirect converters. A direct matrix converter with three-phase input and three-phase output, the switches in a matrix converter must be bi-directional, that is, they must be able to block voltages of either polarity and to conduct current in either direction. This switching strategy permits the highest possible output voltage and reduces the reactive line-side current. Therefore, the power flow through the converter is reversible. Because of its commutation problem and complex control keep it from being broadly utilized in industry.

Unlike the direct matrix converters, the indirect matrix converters has the same functionality, but uses separate input and output sections that are connected through a dc link without storage elements. The design includes a four-quadrant current source rectifier and a voltage source inverter. The input section consists of bi-directional bipolar switches. The commutation strategy can be applied by changing the switching state of the input section while the output section is in a free-wheeling mode. This commutation algorithm is significantly less complexity and higher reliability as compared to a conventional direct matrix converter.

DC link converters: DC Link Converters, also referred to as AC/DC/AC converters, convert an AC input to an AC output with the use of a DC link in the middle. Meaning that the power in the converter is converted to DC from AC with the use of a rectifier, and then it is converted back to AC from DC with the use of an inverter. The end result is an output with a lower voltage and variable (higher or lower) frequency. Due to their wide area of application, the AC/DC/AC converters are the most common contemporary solution. Other advantages to AC/DC/AC converters is that they are stable in overload and no-load conditions, as well as they can be disengaged from a load without damage.

Hybrid matrix converter: Hybrid matrix converters are relatively new for AC/AC converters. These converters combine the AC/DC/AC design with the matrix converter design. Multiple types of hybrid converters have been developed in this new category, an example being a converter that uses uni-directional switches and two converter stages without the dc-link; without the capacitors or inductors needed for a dc-link, the weight and size of the converter is reduced. Two sub-categories exist from the hybrid converters, named hybrid direct matrix converter (HDMC) and hybrid indirect matrix converter (HIMC). HDMC convert the voltage and current in one stage, while the HIMC utilizes separate stages, like the AC/DC/AC converter, but without the use of an intermediate storage element.

Applications: Below is a list of common applications that each converter is used in.

- AC Voltage Controller: Lighting Control; Domestic and Industrial Heating; Speed Control of Fan,Pump or Hoist Drives, Soft Starting of Induction Motors, Static AC Switches (Temperature Control, Transformer Tap Changing, etc.)

- Cycloconverter: High-Power Low-Speed Reversible AC Motor Drives; Constant Frequency

Power Supply with Variable Input Frequency; Controllable VAR Generators for Power Factor Correction; AC System Interties Linking Two Independent Power Systems.

- Matrix Converter: Currently the application of matrix converters are limited due to non-availability of bilateral monolithic switches capable of operating at high frequency, complex control law implementation, commutation and other reasons. With these developments, matrix converters could replace cycloconverters in many areas.

- DC Link: Can be used for individual or multiple load applications of machine building and construction.

Simulations of Power Electronic Systems

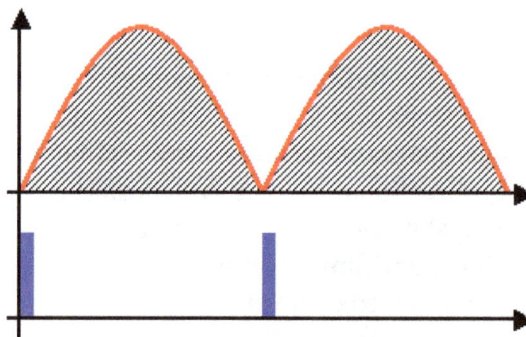

Output voltage of a full-wave rectifier with controlled thyristors

Power electronic circuits are simulated using computer simulation programs such as PLECS, PSIM and MATLAB/simulink. Circuits are simulated before they are produced to test how the circuits respond under certain conditions. Also, creating a simulation is both cheaper and faster than creating a prototype to use for testing.

Applications

Applications of power electronics range in size from a switched mode power supply in an AC adapter, battery chargers, audio amplifiers, fluorescent lamp ballasts, through variable frequency drives and DC motor drives used to operate pumps, fans, and manufacturing machinery, up to gigawatt-scale high voltage direct current power transmission systems used to interconnect electrical grids. Power electronic systems are found in virtually every electronic device. For example:

- DC/DC converters are used in most mobile devices (mobile phones, PDA etc.) to maintain the voltage at a fixed value whatever the voltage level of the battery is. These converters are also used for electronic isolation and power factor correction. A power optimizer is a type of DC/DC converter developed to maximize the energy harvest from solar photovoltaic or wind turbine systems.

- AC/DC converters (rectifiers) are used every time an electronic device is connected to the mains (computer, television etc.). These may simply change AC to DC or can also change the voltage level as part of their operation.

- AC/AC converters are used to change either the voltage level or the frequency (international power adapters, light dimmer). In power distribution networks AC/AC converters may be used to exchange power between utility frequency 50 Hz and 60 Hz power grids.

- DC/AC converters (inverters) are used primarily in UPS or renewable energy systems or emergency lighting systems. Mains power charges the DC battery. If the mains fails, an inverter produces AC electricity at mains voltage from the DC battery. Solar inverter, both smaller string and larger central inverters, as well as solar micro-inverter are used in photovoltaics as a component of a PV system.

Motor drives are found in pumps, blowers, and mill drives for textile, paper, cement and other such facilities. Drives may be used for power conversion and for motion control. For AC motors, applications include variable-frequency drives, motor soft starters and excitation systems.

In hybrid electric vehicles (HEVs), power electronics are used in two formats: series hybrid and parallel hybrid. The difference between a series hybrid and a parallel hybrid is the relationship of the electric motor to the internal combustion engine (ICE). Devices used in electric vehicles consist mostly of dc/dc converters for battery charging and dc/ac converters to power the propulsion motor. Electric trains use power electronic devices to obtain power, as well as for vector control using pulse width modulation (PWM) rectifiers. The trains obtain their power from power lines. Another new usage for power electronics is in elevator systems. These systems may use thyristors, inverters, permanent magnet motors, or various hybrid systems that incorporate PWM systems and standard motors.

Inverters

In general, inverters are utilized in applications requiring direct conversion of electrical energy from DC to AC or indirect conversion from AC to AC. DC to AC conversion is useful for many fields, including power conditioning, harmonic compensation, motor drives, and renewable energy grid-integration.

In power systems it is often desired to eliminate harmonic content found in line currents. VSIs can be used as active power filters to provide this compensation. Based on measured line currents and voltages, a control system determines reference current signals for each phase. This is fed back through an outer loop and subtracted from actual current signals to create current signals for an inner loop to the inverter. These signals then cause the inverter to generate output currents that compensate for the harmonic content. This configuration requires no real power consumption, as it is fully fed by the line; the DC link is simply a capacitor that is kept at a constant voltage by the control system. In this configuration, output currents are in phase with line voltages to produce a unity power factor. Conversely, VAR compensation is possible in a similar configuration where output currents lead line voltages to improve the overall power factor.

In facilities that require energy at all times, such as hospitals and airports, UPS systems are utilized. In a standby system, an inverter is brought online when the normally supplying grid is interrupted. Power is instantaneously drawn from onsite batteries and converted into usable AC voltage by the VSI, until grid power is restored, or until backup generators are brought online. In an online UPS system, a rectifier-DC-link-inverter is used to protect the load from transients and harmonic content. A battery in parallel with the DC-link is kept fully charged by the output in case the grid power is interrupted, while the output of the inverter is fed through a low pass filter to the load. High power quality and independence from disturbances is achieved.

Various AC motor drives have been developed for speed, torque, and position control of AC motors. These drives can be categorized as low-performance or as high-performance, based on whether they are scalar-controlled or vector-controlled, respectively. In scalar-controlled drives, fundamental stator current, or voltage frequency and amplitude, are the only controllable quantities. Therefore, these drives are employed in applications where high quality control is not required, such as fans and compressors. On the other hand, vector-controlled drives allow for instantaneous current and voltage values to be controlled continuously. This high performance is necessary for applications such as elevators and electric cars.

Inverters are also vital to many renewable energy applications. In photovoltaic purposes, the inverter, which is usually a PWM VSI, gets fed by the DC electrical energy output of a photovoltaic module or array. The inverter then converts this into an AC voltage to be interfaced with either a load or the utility grid. Inverters may also be employed in other renewable systems, such as wind turbines. In these applications, the turbine speed usually varies causing changes in voltage frequency and sometimes in the magnitude. In this case, the generated voltage can be rectified and then inverted to stabilize frequency and magnitude.

Smart Grid

A smart grid is a modernized electrical grid that uses information and communications technology to gather and act on information, such as information about the behaviors of suppliers and consumers, in an automated fashion to improve the efficiency, reliability, economics, and sustainability of the production and distribution of electricity.

Electric power generated by wind turbines and hydroelectric turbines by using induction generators can cause variances in the frequency at which power is generated. Power electronic devices are utilized in these systems to convert the generated ac voltages into high-voltage direct current (HVDC). The HVDC power can be more easily converted into three phase power that is coherent with the power associated to the existing power grid. Through these devices, the power delivered by these systems is cleaner and has a higher associated power factor. Wind power systems optimum torque is obtained either through a gearbox or direct drive technologies that can reduce the size of the power electronics device.

Electric power can be generated through photovoltaic cells by using power electronic devices. The produced power is usually then transformed by solar inverters. Inverters are divided into three different types: central, module-integrated and string. Central converters can be connected either in parallel or in series on the DC side of the system. For photovoltaic "farms", a single central converter is used for the entire system. Module-integrated converters are connected in series on either the DC or AC side. Normally several modules are used within a photovoltaic system, since the system requires these converters on both DC and AC terminals. A string converter is used in a system that utilizes photovoltaic cells that are facing different directions. It is used to convert the power generated to each string, or line, in which the photovoltaic cells are interacting.

Grid Voltage Regulation

Power electronics can be used to help utilities adapt to the rapid increase in distributed residential/commercial solar power generation. Germany and parts of Hawaii, California and New Jersey require costly studies to be conducted before approving new solar installations. Relatively small-scale ground- or pole-mounted devices create the potential for a distributed control infrastructure to monitor and manage the flow of power. Traditional electromechanical systems, such as capacitor banks or voltage regulators at substations, can take minutes to adjust voltage and can be distant from the solar installations where the problems originate. If voltage on a neighborhood circuit goes too high, it can endanger utility crews and cause damage to both utility and customer equipment. Further, a grid fault causes photovoltaic generators to shut down immediately, spiking demand for grid power. Smart grid-based regulators are more controllable than far more numerous consumer devices.

In another approach, a group of 16 western utilities called the Western Electric Industry Leaders called for mandatory use of "smart inverters". These devices convert DC to household AC and can also help with power quality. Such devices could eliminate the need for expensive utility equipment upgrades at a much lower total cost.

Commutation Cell

The commutation cell is the basic structure in power electronics. It is composed of an electronic switch (today a high-power semiconductor, not a mechanical switch) and a diode. It was traditionally referred to as a chopper, but since switching power supplies became a major form of power conversion, this new term has become more popular.

The purpose of the commutation cell is to "chop" DC power into square wave alternating current. This is done so that an inductor and a capacitor can be used in an LC circuit to change the voltage. This is in theory a lossless process, and in practice efficiencies above 80-90% are routinely achieved. The output is then usually run through a filter to produce clean DC power. By controlling the on and off times (the duty cycle) of the switch in the commutation cell, the output voltage can be regulated.

This basic principle is the core of most modern power supplies, from tiny DC-DC converters in portable devices to huge switching stations for high voltage DC power transmission.

Connection of Two Power Elements

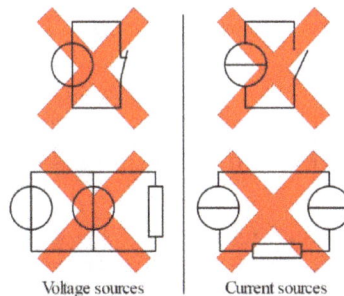

Fig. 1: The different configurations that are impossible: short circuit of a voltage source, current source in an open circuit, two voltage sources in parallel, two current sources in series. Any of these circuits will result in failure of generation of large amounts of heat!

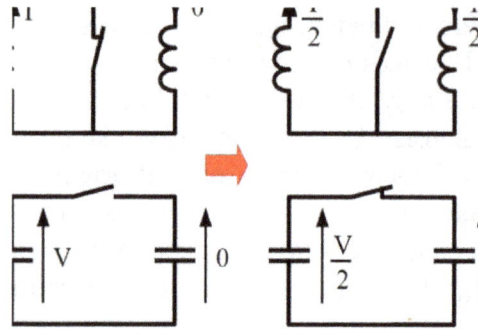

Fig. 2: As with the voltage and current sources, direct energy transfer from a capacitor to another or from an inductor to another should be avoided, as it results in important losses.

A Commutation cell connects two power elements, often referred to as sources, although they can either produce or absorb power.

Some requirements to connect power sources exist. The impossible configurations are listed in figure 1. They are basically:

- a voltage source cannot be shorted, as the short circuit would impose a zero voltage which would contradict the voltage generated by the source;

- in an identical way, a current source cannot be placed in an open circuit;

- two (or more) voltage sources cannot be connected in parallel, as each of them would try to impose the voltage on the circuit;

- two (or more) current sources cannot be connected in series, as each of them would try to impose the current in the loop.

This applies to classical sources (battery, generator), but also to capacitors and inductors: At a small time scale, a capacitor is identical to a voltage source, and an inductor to a current source. Connecting two capacitors with different voltage level in parallel therefore corresponds to connecting two voltage sources, one of the forbidden connections of figure 1.

The figure 2 illustrates the poor efficiency of such connection. One capacitor is charged to a voltage V, and is connected to a capacitor with the same capacity, but discharged.

Before the connection, the energy in the circuit is $E = \frac{1}{2} C \cdot V^2$, and the quantity of charges Q is equal to $C \cdot U$.

After the connection has been made, the quantity of charges is constant, and the total capacitance is $2C$. Therefore the voltage across the capacitances is $\frac{Q}{2C} = \frac{V}{2}$. The energy in the circuit is then $\frac{1}{2}(2C)\left(\frac{V}{2}\right)^2 = \frac{E}{2}$. Therefore half of the energy has been dissipated during the connection.

The same applies with the connections in series of two inductances. The magnetic flux ($\Phi = L \cdot I$) remains constant before and after the commutation. As the total inductance after the commutation is 2L, the current becomes $\frac{I}{2}$. The energy before the commutation is $\frac{1}{2}L \cdot I^2$.

After, it is $\frac{1}{2}L \cdot \left(\frac{I}{2}\right)^2$. Here again, half of the energy is dissipated during the commutation.

As a result, it can be seen that a commutation cell can only connect a voltage source to a current source (and vice versa). However, using inductors and capacitors, it is possible to transform the behaviour of a source: for example two voltage sources can be connected through a converter if it uses an inductor to transfer energy.

The Structure of a Commutation Cell

As told above, a commutation cell must be placed between a voltage source and a current source. Depending on the state of the cell, both sources are either connected, or isolated. When isolated, the current source must be shorted, as it is impossible for a current to be created in an open circuit. The basic schematic of a commutation cell is therefore given in figure 3 (top). It uses two switches with opposite states: In the configuration depicted in figure 3, both sources are isolated, and the current source is shorted. When the top switch is on (and the bottom switch is off), both sources are connected.

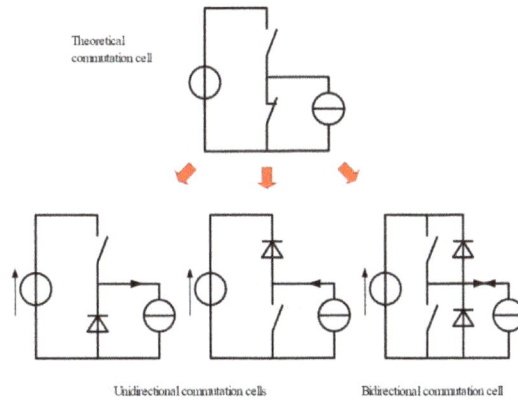

Fig. 3: A commutation cell connects two sources of different nature (current and voltage sources). It theoretically uses two switches, but as they both must be commanded with a perfect synchronization, one of the switches is replaced by a diode in practical applications. This makes the commutation cell unidirectional. A bidirectional commutation cell can be obtained by paralleling two unidirectional ones.

In reality, it is impossible to have a perfect synchronization between the switches. At one point during the commutation, they would be either both on (thus shorting the voltage source) or off (thus leaving the current source in an open circuit). This is why one of the switches has to be replaced by a diode. A diode is a natural commutation device, i.e. its state is controlled by the circuit itself. It will turn on or off at the exact moment it has to. The consequence of using a diode in a commutation cell is that it makes it unidirectional. A bidirectional cell can be built, but it is basically equivalent to two unidirectional cells connected in parallel.

The Commutation Cell in Converters

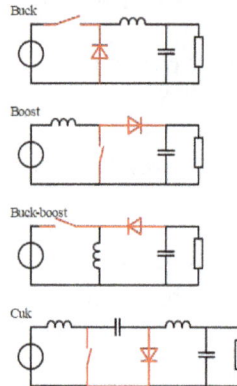

Fig. 4: The commutation cell is present in every switching power supply

The commutation cell can be found in any power electronic converter. Some examples are given in figure 4. As can be seen, a "current source" (actually a loop that contains an inductance) is always connected between the middle point and one of the external connections of the commutation cell, while a voltage source (or a capacitor, or a connection in series of voltage source and capacitor) is always connected to the two external connections.

Switched-Mode Power Supply

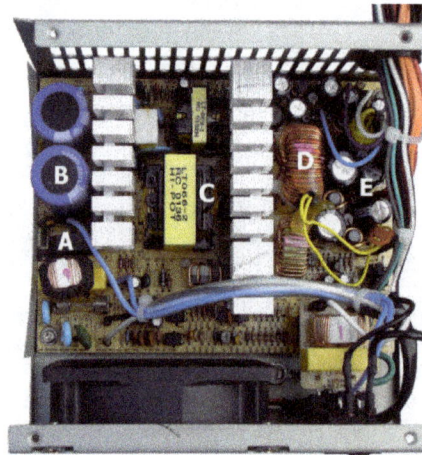

Interior view of an ATX SMPS: below

A: input EMI filtering and bridge rectifier;
B: input filter capacitors;
Between B and C: primary side heat sink; C:
transformer;
Between C and D: secondary side heat sink; D:
output filter coil;
E: output filter capacitors.

The coil and large yellow capacitor below E are additional input filtering components that are mounted directly on the power input connector and are not part of the main circuit board.

A switched-mode power supply (switching-mode power supply, switch-mode power supply, switched power supply, SMPS, or switcher) is an electronic power supply that incorporates a switching regulator to convert electrical power efficiently. Like other power supplies, an SMPS transfers power from a DC or AC source (often mains power), to DC loads, such as a personal computer, while converting voltage and current characteristics. Unlike a linear power supply, the pass transistor of a switching-mode supply continually switches between low-dissipation, full-on and full-off states, and spends very little time in the high dissipation transitions, which minimizes wasted energy. Ideally, a switched-mode power supply dissipates no power. Voltage regulation is achieved by varying the ratio of on-to-off time. In contrast, a linear power supply regulates the output voltage by continually dissipating power in the pass transistor. This higher power conversion efficiency is an important advantage of a switched-mode power supply. Switched-mode power supplies may also be substantially smaller and lighter than a linear supply due to the smaller transformer size and weight.

An adjustable switched-mode power supply for laboratory use

Switching regulators are used as replacements for linear regulators when higher efficiency, smaller size or lighter weight are required. They are, however, more complicated; their switching currents can cause electrical noise problems if not carefully suppressed, and simple designs may have a poor power factor.

History

1836

Induction coils use switches to generate high voltages.

1910

An inductive discharge ignition system invented by Charles F. Kettering and his company Dayton Engineering Laboratories Company (Delco) goes into production for Cadillac. The Kettering ignition system is a mechanically-switched version of a flyback boost converter; the transformer is the ignition coil. Variations of this ignition system were used in all non-diesel internal combustion engines until the 1960s when it was displaced with capacitive discharge ignition systems.

1926

On 23 June, British inventor Philip Ray Coursey applies for a patent in his country and United States, for his "Electrical Condenser". The patent mentions high frequency welding and furnaces, among other uses.

ca 1936

Car radios used electromechanical vibrators to transform the 6 V battery supply to a suitable B+ voltage for the vacuum tubes.

1959

Transistor oscillation and rectifying converter power supply system U.S. Patent 3,040,271 is filed by Joseph E. Murphy and Francis J. Starzec, from General Motors Company

1970

Tektronix starts using High-Efficiency Power Supply in its 7000-series oscilloscopes produced from about 1970 to 1995.

1972

HP-35, Hewlett-Packard's first pocket calculator, is introduced with transistor switching power supply for light-emitting diodes, clocks, timing, ROM, and registers.

1973

Xerox uses switching power supplies in the Alto minicomputer

1977

Apple II is designed with a switching mode power supply. *"Rod Holt was brought in as product engineer and there were several flaws in Apple II that were never publicized. One thing Holt has to his credit is that he created the switching power supply that allowed us to do a very lightweight computer".*

1980

The HP8662A 10 kHz – 1.28 GHz synthesized signal generator went with a switched mode power supply.

Explanation

A linear regulator provides the desired output voltage by dissipating excess power in ohmic losses (e.g., in a resistor or in the collector–emitter region of a pass transistor in its active mode). A linear regulator regulates either output voltage or current by dissipating the excess electric power in the form of heat, and hence its maximum power efficiency is voltage-out/voltage-in since the volt difference is wasted.

In contrast, a switched-mode power supply regulates either output voltage or current by switching ideal storage elements, like inductors and capacitors, into and out of different electrical configurations. Ideal switching elements (e.g., transistors operated outside of their active mode) have no resistance when "closed" and carry no current when "open", and so the converters can theoretically operate with 100% efficiency (i.e., all input power is delivered to the load; no power is wasted as dissipated heat).

For example, if a DC source, an inductor, a switch, and the corresponding electrical ground are placed in series and the switch is driven by a square wave, the peak-to-peak voltage of the waveform measured across the switch can exceed the input voltage from the DC source. This is because the inductor responds to changes in current by inducing its own voltage to counter the change in current, and this voltage adds to the source voltage while the switch is open. If a diode-and-capacitor combination is placed in parallel to the switch, the peak voltage can be stored in the capacitor, and the capacitor can be used as a DC source with an output voltage greater than the DC voltage driving the circuit. This boost converter acts like a step-up transformer for DC signals. A buck–boost converter works in a similar manner, but yields an output voltage which is opposite in polarity to the input voltage. Other buck circuits exist to boost the average output current with a reduction of voltage.

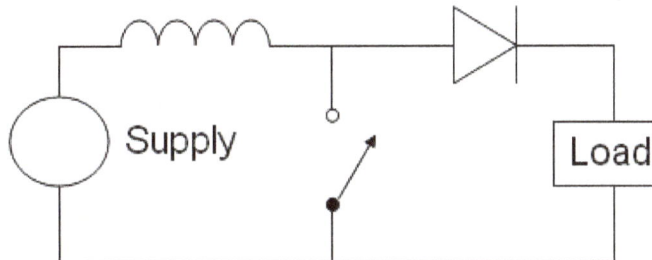

The basic schematic of a boost converter.

In a SMPS, the output current flow depends on the input power signal, the storage elements and circuit topologies used, and also on the pattern used (e.g., pulse-width modulation with an adjustable duty cycle) to drive the switching elements. The spectral density of these switching waveforms has energy concentrated at relatively high frequencies. As such, switching transients and ripple introduced onto the output waveforms can be filtered with a small LC filter.

Advantages and Disadvantages

The main advantage of the switching power supply is greater efficiency than linear regulators because the switching transistor dissipates little power when acting as a switch. Other advantages include smaller size and lighter weight from the elimination of heavy line-frequency transformers, and comparable heat generation. Standby power loss is often much less than transformers.

Disadvantages include greater complexity, the generation of high-amplitude, high-frequency energy that the low-pass filter must block to avoid electromagnetic interference (EMI), a ripple voltage at the switching frequency and the harmonic frequencies thereof.

Very low cost SMPSs may couple electrical switching noise back onto the mains power line, causing interference with A/V equipment connected to the same phase. Non-power-factor-corrected SMPSs also cause harmonic distortion.

SMPS and linear power supply comparison

There are two main types of regulated power supplies available: SMPS and linear. The following table compares linear regulated and unregulated AC-to-DC supplies with switching regulators in general:

Comparison of a linear power supply and a switched-mode power supply			
	Linear power supply	Switching power supply	Notes
Size and weight	Heatsinks for high power linear regulators add size and weight. Transformers, if used, are large due to low operating frequency (mains power frequency is at 50 or 60 Hz); otherwise can be compact due to low component count.	Smaller transformer (if used; else inductor) due to higher operating frequency (typically 50 kHz – 1 MHz). Size and weight of adequate RF shielding may be significant.	A transformer's power handling capacity of given size and weight increases with frequency provided that hysteresis losses can be kept down. Therefore, higher operating frequency means either a higher capacity or smaller transformer.
Output voltage	With transformer used, any voltages available; if transformerless, limited to what can be achieved with a voltage doubler. If unregulated, voltage varies significantly with load.	Any voltages available, limited only by transistor breakdown voltages in many circuits. Voltage varies little with load.	A SMPS can usually cope with wider variation of input before the output voltage changes.
Efficiency, heat, and power dissipation	If regulated: efficiency largely depends on voltage difference between input and output; output voltage is regulated by dissipating excess power as heat resulting in a typical efficiency of 30–40%. If unregulated, transformer iron and copper losses may be the only significant sources of inefficiency.	Output is regulated using duty cycle control; the transistors are switched fully on or fully off, so very little resistive losses between input and the load. The only heat generated is in the non-ideal aspects of the components and quiescent current in the control circuitry.	Switching losses in the transistors (especially in the short part of each cycle when the device is partially on), on-resistance of the switching transistors, equivalent series resistance in the inductor and capacitors, and core losses in the inductor, and rectifier voltage drop contribute to a typical efficiency of 60–70%. However, by optimizing SMPS design (such as choosing the optimal switching frequency, avoiding saturation of inductors, and active rectification), the amount of power loss and heat can be minimized; a good design can have an efficiency of 95%.

Complexity	Unregulated may be simply a diode and capacitor; regulated has a voltage-regulating circuit and a noise-filtering capacitor; usually a simpler circuit (and simpler feedback loop stability criteria) than switched-mode circuits.	Consists of a controller IC, one or several power transistors and diodes as well as a power transformer, inductors, and filter capacitors. Some design complexities present (reducing noise/interference; extra limitations on maximum ratings of transistors at high switching speeds) not found in linear regulator circuits.	In switched-mode mains (AC-to-DC) supplies, multiple voltages can be generated by one transformer core, but that can introduce design/use complications: for example it may place minimum output current restrictions on one output. For this SMPSs have to use duty cycle control. One of the outputs has to be chosen to feed the voltage regulation feedback loop (usually 3.3 V or 5 V loads are more fussy about their supply voltages than the 12 V loads, so this drives the decision as to which feeds the feedback loop. The other outputs usually track the regulated one pretty well). Both need a careful selection of their transformers. Due to the high operating frequencies in SMPSs, the stray inductance and capacitance of the printed circuit board traces become important.
Radio frequency interference	Mild high-frequency interference may be generated by AC rectifier diodes under heavy current loading, while most other supply types produce no high-frequency interference. Some mains hum induction into unshielded cables, problematical for low-signal audio.	EMI/RFI produced due to the current being switched on and off sharply. Therefore, EMI filters and RF shielding are needed to reduce the disruptive interference.	Long wires between the components may reduce the high frequency filter efficiency provided by the capacitors at the inlet and outlet. Stable switching frequency may be important.
Electronic noise at the output terminals	Unregulated PSUs may have a little AC ripple superimposed upon the DC component at twice mains frequency (100–120 Hz). It can cause audible mains hum in audio equipment, brightness ripples or banded distortions in analog security cameras.	Noisier due to the switching frequency of the SMPS. An unfiltered output may cause glitches in digital circuits or noise in audio circuits.	This can be suppressed with capacitors and other filtering circuitry in the output stage. With a switched mode PSU the switching frequency can be chosen to keep the noise out of the circuits working frequency band (e.g., for audio systems above the range of human hearing)
Electronic noise at the input terminals	Causes harmonic distortion to the input AC, but relatively little or no high frequency noise.	Very low cost SMPS may couple electrical switching noise back onto the mains power line, causing interference with A/V equipment connected to the same phase. Non power-factor-corrected SMPSs also cause harmonic distortion.	This can be prevented if a (properly earthed) EMI/RFI filter is connected between the input terminals and the bridge rectifier.

Acoustic noise	Faint, usually inaudible mains hum, usually due to vibration of windings in the transformer or magnetostriction.	Usually inaudible to most humans, unless they have a fan or are unloaded/malfunctioning, or use a switching frequency within the audio range, or the laminations of the coil vibrate at a subharmonic of the operating frequency.	The operating frequency of an unloaded SMPS is sometimes in the audible human range, and may sound subjectively quite loud for people whose hearing is very sensitive to the relevant frequency range.
Power factor	Low for a regulated supply because current is drawn from the mains at the peaks of the voltage sinusoid, unless a choke-input or resistor-input circuit follows the rectifier (now rare).	Ranging from very low to medium since a simple SMPS without PFC draws current spikes at the peaks of the AC sinusoid.	Active/passive power factor correction in the SMPS can offset this problem and are even required by some electric regulation authorities, particularly in the EU. The internal resistance of low-power transformers in linear power supplies usually limits the peak current each cycle and thus gives a better power factor than many switched-mode power supplies that directly rectify the mains with little series resistance.
Inrush current	Large current when mains-powered linear power supply equipment is switched on until magnetic flux of transformer stabilises and capacitors charge completely, unless a slow-start circuit is used.	Extremely large peak "in-rush" surge current limited only by the impedance of the input supply and any series resistance to the filter capacitors.	Empty filter capacitors initially draw large amounts of current as they charge up, with larger capacitors drawing larger amounts of peak current. Being many times above the normal operating current, this greatly stresses components subject to the surge, complicates fuse selection to avoid nuisance blowing and may cause problems with equipment employing overcurrent protection such as uninterruptible power supplies. Mitigated by use of a suitable soft-start circuit or series resistor.
Risk of electric shock	Supplies with transformers isolate the incoming power supply from the powered device and so allow metalwork of the enclosure to be grounded safely. Dangerous if primary/secondary insulation breaks down, unlikely with reasonable design. Transformerless mains-operated supply dangerous. In both linear and switch-mode the mains, and possibly the output voltages, are hazardous and must be well-isolated.	Common rail of equipment (including casing) is energized to half the mains voltage, but at high impedance, unless equipment is earthed/grounded or doesn't contain EMI/RFI filtering at the input terminals.	Due to regulations concerning EMI/RFI radiation, many SMPS contain EMI/RFI filtering at the input stage before the bridge rectifier consisting of capacitors and inductors. Two capacitors are connected in series with the Live and Neutral rails with the Earth connection in between the two capacitors. This forms a capacitive divider that energizes the common rail at half mains voltage. Its high impedance current source can provide a tingling or a 'bite' to the operator or can be exploited to light an Earth Fault LED. However, this current may cause nuisance tripping on the most sensitive residual-current devices.

		Can fail so as to make output voltage very high. Stress on capacitors may cause them to explode. Can in some cases destroy input stages in amplifiers if floating voltage exceeds transistor base-emitter breakdown voltage, causing the transistor's gain to drop and noise levels to increase. Mitigated by good failsafe design. Failure of a component in the SMPS itself can cause further damage to other PSU components; can be difficult to troubleshoot.	The floating voltage is caused by capacitors bridging the primary and secondary sides of the power supply. Connection to earthed equipment will cause a momentary (and potentially destructive) spike in current at the connector as the voltage at the secondary side of the capacitor equalizes to earth potential.
Risk of equipment damage	Very low, unless a short occurs between the primary and secondary windings or the regulator fails by shorting internally.		

Theory of Operation

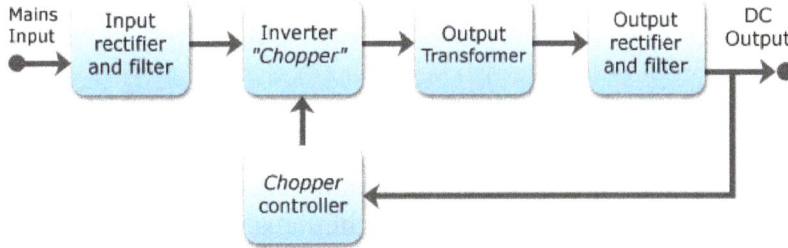

Block diagram of a mains operated AC/DC SMPS with output voltage regulation

Input Rectifier Stage

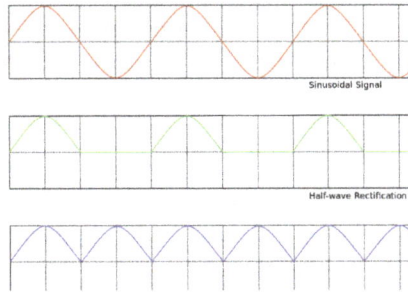

AC, half-wave and full-wave rectified signals.

If the SMPS has an AC input, then the first stage is to convert the input to DC. This is called *rectification*. A SMPS with a DC input does not require this stage. In some power supplies (mostly computer ATX power supplies), the rectifier circuit can be configured as a voltage doubler by the addition of a switch operated either manually or automatically. This feature permits operation from power sources that are normally at 115 V or at 230 V. The rectifier produces an unregulated DC voltage which is then sent to a large filter capacitor. The current drawn from the mains supply by this rectifier circuit occurs in short pulses around the AC voltage peaks. These pulses have significant high frequency energy which reduces the power factor. To correct for this, many newer SMPS will use a special PFC circuit to make the input current follow the sinusoidal shape of the AC input voltage, correcting the power factor. Power supplies that use Active PFC usually are auto-ranging, supporting input voltages from ~100 VAC – 250 VAC, with no input voltage selector switch.

An SMPS designed for AC input can usually be run from a DC supply, because the DC would pass through the rectifier unchanged. If the power supply is designed for 115 VAC and has no voltage selector switch, the required DC voltage would be 163 VDC (115 × √2). This type of use may be harmful to the rectifier stage, however, as it will only use half of diodes in the rectifier for the full load. This could possibly result in overheating of these components, causing them to fail prematurely. On the other hand, if the power supply has a voltage selector switch, based on the Delon circuit, for 115/230V (computer ATX power supplies typically are in this category), the selector switch would have to be put in the 230 V position, and the required voltage would be 325 VDC (230 × √2). The diodes in this type of power supply will handle the DC current just fine because they are rated to handle double the nominal input current when operated in the 115 V mode, due to the operation of the voltage doubler. This is because the doubler, when in operation, uses only half of the bridge rectifier and runs twice as much current through it. It is uncertain how an Auto-ranging/ Active-PFC type power supply would react to being powered by DC.

Inverter Stage

The inverter stage converts DC, whether directly from the input or from the rectifier stage described above, to AC by running it through a power oscillator, whose output transformer is very small with few windings at a frequency of tens or hundreds of kilohertz. The frequency is usually chosen to be above 20 kHz, to make it inaudible to humans. The switching is implemented as a multistage (to achieve high gain) MOSFET amplifier. MOSFETs are a type of transistor with a low on-resistance and a high current-handling capacity.

Voltage Converter and Output Rectifier

If the output is required to be isolated from the input, as is usually the case in mains power supplies, the inverted AC is used to drive the primary winding of a high-frequency transformer. This converts the voltage up or down to the required output level on its secondary winding. The output transformer in the block diagram serves this purpose.

If a DC output is required, the AC output from the transformer is rectified. For output voltages above ten volts or so, ordinary silicon diodes are commonly used. For lower voltages, Schottky diodes are commonly used as the rectifier elements; they have the advantages of faster recovery times than silicon diodes (allowing low-loss operation at higher frequencies) and a lower voltage drop when conducting. For even lower output voltages, MOSFETs may be used as synchronous rectifiers; compared to Schottky diodes, these have even lower conducting state voltage drops.

The rectified output is then smoothed by a filter consisting of inductors and capacitors. For higher switching frequencies, components with lower capacitance and inductance are needed.

Simpler, non-isolated power supplies contain an inductor instead of a transformer. This type includes *boost converters, buck converters*, and the *buck-boost converters*. These belong to the simplest class of single input, single output converters which use one inductor and one active switch. The buck converter reduces the input voltage in direct proportion to the ratio of conductive time to the total switching period, called the duty cycle. For example an ideal buck converter with a 10 V input operating at a 50% duty cycle will produce an average output voltage of 5 V. A feedback control loop is employed to regulate the output voltage by varying the duty cycle to compensate for variations in input voltage. The output voltage of a boost converter is always greater than the input voltage and the buck-boost output voltage is inverted but can be greater than, equal to, or less than the magnitude of its input voltage. There are many variations and extensions to this class of converters but these three form the basis of almost all isolated and non-isolated DC to DC converters. By adding a second inductor the Ćuk and SEPIC converters can be implemented, or, by adding additional active switches, various bridge converters can be realized.

Other types of SMPSs use a capacitor-diode voltage multiplier instead of inductors and transformers. These are mostly used for generating high voltages at low currents (*Cockcroft-Walton generator*). The low voltage variant is called charge pump.

Regulation

A feedback circuit monitors the output voltage and compares it with a reference voltage, as shown in the block diagram above. Depending on design and safety requirements, the controller may contain an isolation mechanism (such as an opto-coupler) to isolate it from the DC output. Switching supplies in computers, TVs and VCRs have these opto-couplers to tightly control the output voltage.

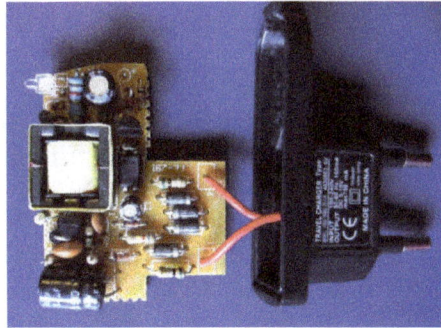

This charger for a small device such as a mobile phone is a simple off-line switching power supply with a European plug.

Open-loop regulators do not have a feedback circuit. Instead, they rely on feeding a constant voltage to the input of the transformer or inductor, and assume that the output will be correct. Regulated designs compensate for the impedance of the transformer or coil. Monopolar designs also compensate for the magnetic hysteresis of the core.

The feedback circuit needs power to run before it can generate power, so an additional non-switching power-supply for stand-by is added.

Transformer Design

Any switched-mode power supply that gets its power from an AC power line (called an "off-line" converter) requires a transformer for galvanic isolation. Some DC-to-DC converters may also include a transformer, although isolation may not be critical in these cases. SMPS transformers run at high frequency. Most of the cost savings (and space savings) in off-line power supplies result from the smaller size of the high frequency transformer compared to the 50/60 Hz transformers formerly used. There are additional design tradeoffs.

The terminal voltage of a transformer is proportional to the product of the core area, magnetic flux, and frequency. By using a much higher frequency, the core area (and so the mass of the core) can be greatly reduced. However, core losses increase at higher frequencies. Cores generally use ferrite material which has a low loss at the high frequencies and high flux densities used. The laminated iron cores of lower-frequency (<400 Hz) transformers would be unacceptably lossy at switching frequencies of a few kilohertz. Also, more energy is lost during transitions of the switching semi-conductor at higher frequencies. Furthermore, more attention to the physical layout of the circuit board is required as parasitics become more significant, and the amount of electromagnetic inter-ference will be more pronounced.

Copper Loss

At low frequencies (such as the line frequency of 50 or 60 Hz), designers can usually ignore the skin effect. For these frequencies, the skin effect is only significant when the conductors are large, more than 0.3 inches (7.6 mm) in diameter.

Switching power supplies must pay more attention to the skin effect because it is a source of power loss. At 500 kHz, the skin depth in copper is about 0.003 inches (0.076 mm) – a dimension small-er than the typical wires used in a power supply. The effective resistance of conductors increases, because current concentrates near the surface of the conductor and the inner portion carries less current than at low frequencies.

The skin effect is exacerbated by the harmonics present in the high speed PWM switching wave-forms. The appropriate skin depth is not just the depth at the fundamental, but also the skin depths at the harmonics.

In addition to the skin effect, there is also a proximity effect, which is another source of power loss.

Power Factor

Simple off-line switched mode power supplies incorporate a simple full-wave rectifier connected to a large energy storing capacitor. Such SMPSs draw current from the AC line in short pulses when the mains instantaneous voltage exceeds the voltage across this capacitor. During the remaining portion of the AC cycle the capacitor provides energy to the power supply.

As a result, the input current of such basic switched mode power supplies has high harmonic content and relatively low power factor. This creates extra load on utility lines, increases heating of building wiring, the utility transformers, and standard AC electric motors, and may cause sta-bility problems in some applications such as in emergency generator systems or aircraft gener-ators. Harmonics can be removed by filtering, but the filters are expensive. Unlike displacement power factor created by linear inductive or capacitive loads, this distortion cannot be corrected by addition of a single linear component. Additional circuits are required to counteract the effect of the brief current pulses. Putting a current regulated boost chopper stage after the off-line rectifier (to charge the storage capacitor) can correct the power factor, but increases the complexity and cost.

In 2001, the European Union put into effect the standard IEC/EN61000-3-2 to set limits on the harmonics of the AC input current up to the 40th harmonic for equipment above 75 W. The standard defines four classes of equipment depending on its type and current waveform. The most rigorous limits (class D) are established for personal computers, computer monitors, and TV receivers. To comply with these requirements, modern switched-mode power supplies normally include an additional power factor correction (PFC) stage.

Types

Switched-mode power supplies can be classified according to the circuit topology. The most important distinction is between isolated converters and non-isolated ones.

Non-isolated topologies

Non-isolated converters are simplest, with the three basic types using a single inductor for energy storage. In the voltage relation column, D is the duty cycle of the converter, and can vary from 0 to 1. The input voltage (V_1) is assumed to be greater than zero; if it is negative, for consistency, negate the output voltage (V_2).

Type	Typical Power [W]	Relative cost	Energy storage	Voltage relation	Features
Buck	0–1,000	1.0	Single inductor	$0 \leq \text{Out} \leq \text{In},$ $V_2 = DV_1$	Current is continuous at output.
Boost	0–5,000	1.0	Single inductor	$\text{Out} \geq \text{In},$ $V_2 = \dfrac{1}{1-D}V_1$	Current is continuous at input.
Buck-boost	0–150	1.0	Single inductor	$\text{Out} \leq 0,$ $V_2 = -\dfrac{D}{1-D}V_1$	Current is dis-continuous at both input and output.
Split-pi (or, boost-buck)	0–4,500	>2.0	Two inductors and three capacitors	Up or down	Bidirectional power control; in or out
Ćuk			Capacitor and two inductors	Any inverted, $V_2 = -\dfrac{D}{1-D}V_1$	Current is continuous at input and output

SEPIC			Capacitor and two inductors	Any, $V_2 = \dfrac{D}{1-D} V_1$	Current is continuous at input
Zeta			Capacitor and two inductors	Any, $V_2 = \dfrac{D}{1-D} V_1$	Current is continuous at output
Charge pump / Switched capacitor			Capacitors only		No magnetic energy storage is needed to achieve conversion, however high efficiency power processing is normally limited to a discrete set of conversion ratios.

When equipment is human-accessible, voltage and power limits of <=42.4 V peak/60 V dc and 250 VA apply for Safety Certification (UL, CSA, VDE approval).

The buck, boost, and buck-boost topologies are all strongly related. Input, output and ground come together at one point. One of the three passes through an inductor on the way, while the other two pass through switches. One of the two switches must be active (e.g., a transistor), while the other can be a diode. Sometimes, the topology can be changed simply by re-labeling the connections. A 12 V input, 5 V output buck converter can be converted to a 7 V input, −5 V output buck-boost by grounding the *output* and taking the output from the *ground* pin.

Likewise, SEPIC and Zeta converters are both minor rearrangements of the Ćuk converter.

The Neutral Point Clamped (NPC) topology is used in power supplies and active filters and is mentioned here for completeness.

Switchers become less efficient as duty cycles become extremely short. For large voltage changes, a transformer (isolated) topology may be better.

Isolated Topologies

All isolated topologies include a transformer, and thus can produce an output of higher or lower voltage than the input by adjusting the turns ratio. For some topologies, multiple windings can be placed on the transformer to produce multiple output voltages. Some converters use the transformer for energy storage, while others use a separate inductor.

Type	Power [W]	Relative cost	Input range [V]	Energy storage	Features
Flyback	0–250	1.0	5–600	Mutual Inductors	Isolated form of the buck-boost converter.[1]

Ringing choke converter (RCC)	0–150	1.0	5–600	Transformer	Low-cost self-oscillating flyback variant.
Half-forward	0–250	1.2	5–500	Inductor	
Forward²	100-200		60–200	Inductor	Isolated form of buck converter
Resonant forward	0–60	1.0	60–400	Inductor and capacitor	Single rail input, unregulated output, high efficiency, low EMI.
Push-pull	100–1,000	1.75	50–1,000	Inductor	
Half-bridge	0–2,000	1.9	50–1,000	Inductor	
Full-bridge	400–5,000	>2.0	50–1,000	Inductor	Very efficient use of transformer, used for highest powers.
Resonant, zero voltage switched	>1,000	>2.0		Inductor and capacitor	
Isolated Ćuk				Two capacitors and two inductors	

Zero voltage switched mode power supplies require only small heatsinks as little energy is lost as heat. This allows them to be small. This ZVS can deliver more than 1 kilowatt. Transformer is not shown.

- Flyback converter logarithmic control loop behavior might be harder to control than other types.
- The forward converter has several variants, varying in how the transformer is "reset" to zero magnetic flux every cycle.

Quasi-Resonant Zero-Current/Zero-Voltage Switch

In a quasi-resonant zero-current/zero-voltage switch (ZCS/ZVS) "each switch cycle delivers a quantized 'packet' of energy to the converter output, and switch turn-on and turn-off occurs at zero current and voltage, resulting in an essentially lossless switch." Quasi-resonant switching, also known as *valley switching*, reduces EMI in the power supply by two methods:

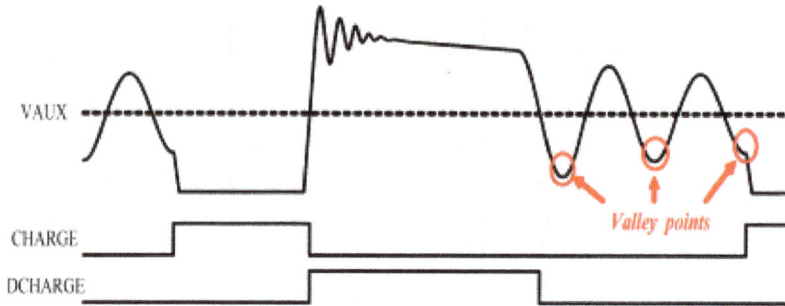

Quasi-resonant switching switches when the voltage is at a minimum and a valley is detected

By switching the bipolar switch when the voltage is at a minimum (in the valley) to minimize the hard switching effect that causes EMI.

By switching when a valley is detected, rather than at a fixed frequency, introduces a natural frequency jitter that spreads the RF emissions spectrum and reduces overall EMI.

Efficiency and EMI

Higher input voltage and synchronous rectification mode makes the conversion process more efficient. The power consumption of the controller also has to be taken into account. Higher switching frequency allows component sizes to be shrunk, but can produce more RFI. A resonant forward converter produces the lowest EMI of any SMPS approach because it uses a soft-switching resonant waveform compared with conventional hard switching.

Failure Modes

Power supplies which use capacitors suffering from the capacitor plague may experience premature failure when the capacitance drops to 4% of the original value. This usually causes the switching semiconductor to fail in a conductive way. That may expose connected loads to the full input volt and current, and precipitate wild oscillations in output.

Failure of the switching transistor is common. Due to the large switching voltages this transistor must handle (around 325 V for a 230 V_{AC} mains supply), these transistors often short out, in turn immediately blowing the main internal power fuse.

Precautions

The main filter capacitor will often store up to 325 volts long after the power cord has been removed from the wall. Not all power supplies contain a small "bleeder" resistor to slowly discharge this capacitor. Any contact with this capacitor may result in a severe electrical shock.

The primary and secondary sides may be connected with a capacitor to reduce EMI and compensate for various capacitive couplings in the converter circuit, where the transformer is one. This may result in electric shock in some cases. The current flowing from line or neutral through a 2 kΩ resistor to any accessible part must, according to IEC 60950, be less than 250 µA for IT equipment.

Applications

Switched Mode Mobile Phone Charger

A 450 Watt SMPS for use in personal computers with the power input, fan, and output cords visible

Switched-mode power supply units (PSUs) in domestic products such as personal computers often have universal inputs, meaning that they can accept power from mains supplies throughout the world, although a manual voltage range switch may be required. Switch-mode power supplies can tolerate a wide range of power frequencies and voltages.

Due to their high volumes mobile phone chargers have always been particularly cost sensitive. The first chargers were linear power supplies but they quickly moved to the cost effective ringing choke converter (RCC) SMPS topology, when new levels of efficiency were required. Recently, the demand for even lower no-load power requirements in the application has meant that flyback topology is being used more widely; primary side sensing flyback controllers are also helping to cut the bill of materials (BOM) by removing secondary-side sensing components such as optocouplers.

Switched-mode power supplies are used for DC to DC conversion as well. In automobiles where heavy vehicles use a nominal 24 V_{DC} cranking supply, 12V for accessories may be furnished through a DC/DC switch-mode supply. This has the advantage over tapping the battery at the 12V position (using half the cells) that all the 12V load is evenly divided over all cells of the 24V battery. In industrial settings such as telecommunications racks, bulk power may be distributed at a low DC voltage (from a battery back up system, for example) and individual equipment items will have DC/DC switched-mode converters to supply whatever voltages are needed.

Terminology

The term switchmode was widely used until Motorola claimed ownership of the trademark SWITCHMODE, for products aimed at the switching-mode power supply market, and started to enforce their trademark. *Switching-mode power supply*, *switching power supply*, and *switching regulator* refer to this type of power supply.

References

- Editor: Semikron, Authors: Dr. Ulrich Nicolai, Dr. Tobias Reimann, Prof. Jürgen Petzoldt, Josef Lutz: Application Manual IGBT- and MOSFET-power modules, 1. edition, ISLE Verlag, 1998, ISBN 3-932633-24-5.

- R. W. Erickson, D. Maksimovic, Fundamentals of Power Electronics, 2nd Ed., Springer, 2001, ISBN 0-7923-7270-0 [1]

- Arendt Wintrich, Ulrich Nicolai, Werner Tursky, Tobias Reimann (2010) (in German), [PDF-Version Applikationshandbuch 2010] (2. ed.), ISLE Verlag, ISBN 978-3-938843-56-7.

- Arendt Wintrich, Ulrich Nicolai, Werner Tursky, Tobias Reimann (2011) (in German), [PDF-Version Application Manual 2011] (2. ed.), ISLE Verlag, ISBN 978-3-938843-66-6

- Pressman, Abraham I. (1998), Switching Power Supply Design (2nd ed.), McGraw-Hill, ISBN 0-07-052236-7

Mechanisms and Methods to Produce Electric Power

The generation of electric power has been the focal point of civilization and the history of electricity generation can be traced back to as long ago as 6th century BC. This chapter briefs the reader about mechanisms and methods of electric energy generation like wind turbine, hydroelectricity, induction generator, solar cell, tidal stream generator and windbelt. The section on wind turbine also talks about the modification of wind lens. The content is tailored to provide elaborate information on the devices and their operations and applications.

Wind Turbine

Offshore wind farm, using 5 MW turbines REpower 5M in the North Sea off the coast of Belgium.

A wind turbine is a device that converts the wind's kinetic energy into electrical power. The term appears to have been adopted from hydroelectric technology (rotary propeller). The technical description of a wind turbine is aerofoil-powered generator.

As a result of over a millennium of windmill development and modern engineering, today's wind turbines are manufactured in a wide range of vertical and horizontal axis types. The smallest turbines are used for applications such as battery charging for auxiliary power for boats or caravans or to power traffic warning signs. Slightly larger turbines can be used for making contributions to a domestic power supply while selling unused power back to the utility supplier via the electrical grid. Arrays of large turbines, known as wind farms, are becoming an increasingly important source of renewable energy and are used by many countries as part of a strategy to reduce their reliance on fossil fuels.

History

James Blyth's electricity-generating wind turbine, photographed in 1891

Windmills were used in Persia (present-day Iran) about 500-900 A.D. The windwheel of Hero of Alexandria marks one of the first known instances of wind powering a machine in history. However, the first known practical windmills were built in Sistan, an Eastern province of Iran, from the 7th century. These "Panemone" were vertical axle windmills, which had long vertical drive shafts with rectangular blades. Made of six to twelve sails covered in reed matting or cloth material, these windmills were used to grind grain or draw up water, and were used in the gristmilling and sugarcane industries.

Windmills first appeared in Europe during the Middle Ages. The first historical records of their use in England date to the 11th or 12th centuries and there are reports of German crusaders taking their windmill-making skills to Syria around 1190. By the 14th century, Dutch windmills were in use to drain areas of the Rhine delta. Advanced wind mills were described by Croatian inventor Fausto Veranzio. In his book Machinae Novae (1595) he described vertical axis wind turbines with curved or V-shaped blades.

The first electricity-generating wind turbine was a battery charging machine installed in July 1887 by Scottish academic James Blyth to light his holiday home in Marykirk, Scotland. Some months later American inventor Charles F. Brush was able to build the first automatically operated wind turbine after consulting local University professors and colleagues Jacob S. Gibbs and Brinsley Coleberd and successfully getting the blueprints peer-reviewed for electricity production in Cleveland, Ohio. Although Blyth's turbine was considered uneconomical in the United Kingdom electricity generation by wind turbines was more cost effective in countries with widely scattered populations.

In Denmark by 1900, there were about 2500 windmills for mechanical loads such as pumps and mills, producing an estimated combined peak power of about 30 MW. The largest machines were on 24-meter (79 ft) towers with four-bladed 23-meter (75 ft) diameter rotors. By 1908 there were 72 wind-driven electric generators operating in the United States from 5 kW to 25 kW. Around the time of World War I, American windmill makers were producing 100,000 farm windmills each year, mostly for water-pumping.

The first automatically operated wind turbine, built in Cleveland in 1887 by Charles F. Brush. It was 60 feet (18 m) tall, weighed 4 tons (3.6 metric tonnes) and powered a 12 kW generator.

By the 1930s, wind generators for electricity were common on farms, mostly in the United States where distribution systems had not yet been installed. In this period, high-tensile steel was cheap, and the generators were placed atop prefabricated open steel lattice towers.

A forerunner of modern horizontal-axis wind generators was in service at Yalta, USSR in 1931. This was a 100 kW generator on a 30-meter (98 ft) tower, connected to the local 6.3 kV distribution system. It was reported to have an annual capacity factor of 32 percent, not much different from current wind machines.

In the autumn of 1941, the first megawatt-class wind turbine was synchronized to a utility grid in Vermont. The Smith-Putnam wind turbine only ran for 1,100 hours before suffering a critical failure. The unit was not repaired, because of shortage of materials during the war.

The first utility grid-connected wind turbine to operate in the UK was built by John Brown & Company in 1951 in the Orkney Islands.

Despite these diverse developments, developments in fossil fuel systems almost entirely eliminated any wind turbine systems larger than supermicro size. In the early 1970s, however, anti-nuclear protests in Denmark spurred artisan mechanics to develop microturbines of 22 kW. Organizing owners into associations and co-operatives lead to the lobbying of the government and utilities and provided incentives for larger turbines throughout the 1980s and later. Local activists in Germany, nascent turbine manufacturers in Spain, and large investors in the United States in the early 1990s then lobbied for policies that stimulated the industry in those countries. Later companies formed in India and China. As of 2012, Danish company Vestas is the world's biggest wind-turbine manufacturer.

Resources

A quantitative measure of the wind energy available at any location is called the Wind Power Density (WPD). It is a calculation of the mean annual power available per square meter of swept area of a turbine, and is tabulated for different heights above ground. Calculation of wind power density includes the effect of wind velocity and air density. Color-coded maps are prepared for a particular area described, for example, as "Mean Annual Power Density at 50 Metres". In the United States, the results of the above calculation are included in an index developed by the National Renewable

Energy Laboratory and referred to as "NREL CLASS". The larger the WPD, the higher it is rated by class. Classes range from Class 1 (200 watts per square meter or less at 50 m altitude) to Class 7 (800 to 2000 watts per square m). Commercial wind farms generally are sited in Class 3 or higher areas, although isolated points in an otherwise Class 1 area may be practical to exploit.

Nordex N117/2400 in Germany, a modern low-wind turbine.

Wind turbines at the Jepirachí Eolian Park in La Guajira, Colombia.

Wind turbines are classified by the wind speed they are designed for, from class I to class IV, with A or B referring to the turbulence.

Class	Avg Wind Speed (m/s)	Turbulence
IA	10	18%
IB	10	16%
IIA	8.5	18%
IIB	8.5	16%
IIIA	7.5	18%

IIIB	7.5	16%
IVA	6	18%
IVB	6	16%

Efficiency

Not all the energy of blowing wind can be used, but some small wind turbines are designed to work at low wind speeds.

Conservation of mass requires that the amount of air entering and exiting a turbine must be equal. Accordingly, Betz's law gives the maximal achievable extraction of wind power by a wind turbine as 16/27 (59.3%) of the total kinetic energy of the air flowing through the turbine.

The maximum theoretical power output of a wind machine is thus 0.59 times the kinetic energy of the air passing through the effective disk area of the machine. If the effective area of the disk is A, and the wind velocity v, the maximum theoretical power output P is:

$$P = 0.59 \frac{1}{2} \rho v^3 A$$

where ρ is air density

As wind is free (no fuel cost), wind-to-rotor efficiency (including rotor blade friction and drag) is one of many aspects impacting the final price of wind power. Further inefficiencies, such as gearbox losses, generator and converter losses, reduce the power delivered by a wind turbine. To protect components from undue wear, extracted power is held constant above the rated operating speed as theoretical power increases at the cube of wind speed, further reducing theoretical efficiency. In 2001, commercial utility-connected turbines deliver 75% to 80% of the Betz limit of power extractable from the wind, at rated operating speed.

Efficiency can decrease slightly over time due to wear. Analysis of 3128 wind turbines older than 10 years in Denmark showed that half of the turbines had no decrease, while the other half saw a production decrease of 1.2% per year.

Types

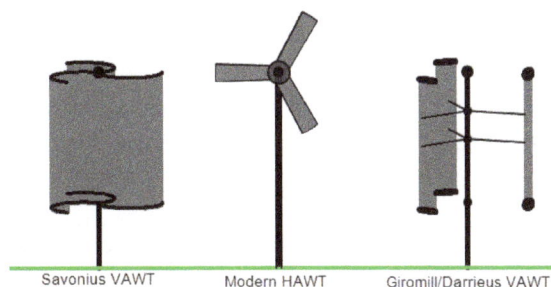

Savonius VAWT   Modern HAWT   Giromill/Darrieus VAWT

The three primary types: VAWT Savonius, HAWT towered; VAWT Darrieus as they appear in operation

Wind turbines can rotate about either a horizontal or a vertical axis, the former being both older and more common. They can also include blades (transparent or not) or be bladeless.

Horizontal Axis

Components of a horizontal axis wind turbine (gearbox, rotor shaft and brake assembly) being lifted into position

Horizontal-axis wind turbines (HAWT) have the main rotor shaft and electrical generator at the top of a tower, and must be pointed into the wind. Small turbines are pointed by a simple wind vane, while large turbines generally use a wind sensor coupled with a servo motor. Most have a gearbox, which turns the slow rotation of the blades into a quicker rotation that is more suitable to drive an electrical generator.

A turbine blade convoy passing through Edenfield, UK

Since a tower produces turbulence behind it, the turbine is usually positioned upwind of its supporting tower. Turbine blades are made stiff to prevent the blades from being pushed into the tower by high winds. Additionally, the blades are placed a considerable distance in front of the tower and are sometimes tilted forward into the wind a small amount.

Downwind machines have been built, despite the problem of turbulence (mast wake), because they don't need an additional mechanism for keeping them in line with the wind, and because in high winds the blades can be allowed to bend which reduces their swept area and thus their wind resistance. Since cyclical (that is repetitive) turbulence may lead to fatigue failures, most HAWTs are of upwind design.

Turbines used in wind farms for commercial production of electric power are usually three-bladed and pointed into the wind by computer-controlled motors. These have high tip speeds of over 320 km/h (200 mph), high efficiency, and low torque ripple, which contribute to good reliability.

The blades are usually colored white for daytime visibility by aircraft and range in length from 20 to 40 meters (66 to 131 ft) or more. The tubular steel towers range from 60 to 90 meters (200 to 300 ft) tall.

The blades rotate at 10 to 22 revolutions per minute. At 22 rotations per minute the tip speed exceeds 90 meters per second (300 ft/s). A gear box is commonly used for stepping up the speed of the generator, although designs may also use direct drive of an annular generator. Some models operate at constant speed, but more energy can be collected by variable-speed turbines which use a solid-state power converter to interface to the transmission system. All turbines are equipped with protective features to avoid damage at high wind speeds, by feathering the blades into the wind which ceases their rotation, supplemented by brakes.

Year by year the size and height of turbines increase. Offshore wind turbines are built up to 8MW today and have a blade length up to 80m. Onshore wind turbines are installed in low wind speed areas and getting higher and higher towers. Usual towers of multi megawatt turbines have a height of 70 m to 120 m and in extremes up to 160 m, with blade tip speeds reaching 80 m/s to 90 m/s. Higher tip speeds means more noise and blade erosion.

Vertical Axis Design

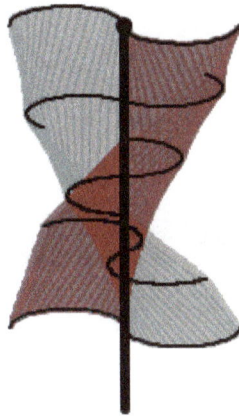

A vertical axis Twisted Savonius type turbine.

Vertical-axis wind turbines (or VAWTs) have the main rotor shaft arranged vertically. One advantage of this arrangement is that the turbine does not need to be pointed into the wind to be effective, which is an advantage on a site where the wind direction is highly variable. It is also an advantage when the turbine is integrated into a building because it is inherently less steerable. Also, the generator and gearbox can be placed near the ground, using a direct drive from the rotor assembly to the ground-based gearbox, improving accessibility for maintenance.

The key disadvantages include the relatively low rotational speed with the consequential higher torque and hence higher cost of the drive train, the inherently lower power coefficient, the 360-degree rotation of the aerofoil within the wind flow during each cycle and hence the highly dynamic loading on the blade, the pulsating torque generated by some rotor designs on the drive train, and the difficulty of modelling the wind flow accurately and hence the challenges of analysing and designing the rotor prior to fabricating a prototype.

When a turbine is mounted on a rooftop the building generally redirects wind over the roof and this can double the wind speed at the turbine. If the height of a rooftop mounted turbine tower is approximately 50% of the building height it is near the optimum for maximum wind energy and minimum wind turbulence. Wind speeds within the built environment are generally much lower than at exposed rural sites, noise may be a concern and an existing structure may not adequately resist the additional stress.

Subtypes of the vertical axis design include:

Offshore Horizontal Axis Wind Turbines (HAWTs) at Scroby Sands Wind Farm, UK

Onshore Horizontal Axis Wind Turbines in Zhangjiakou, China

Darrieus Wind Turbine

"Eggbeater" turbines, or Darrieus turbines, were named after the French inventor, Georges Darrieus. They have good efficiency, but produce large torque ripple and cyclical stress on the tower, which contributes to poor reliability. They also generally require some external power source, or an additional Savonius rotor to start turning, because the starting torque is very low. The torque ripple is reduced by using three or more blades which results in greater solidity of the rotor. Solidity is measured by blade area divided by the rotor area. Newer Darrieus type turbines are not held up by guy-wires but have an external superstructure connected to the top bearing.

Giromill

A subtype of Darrieus turbine with straight, as opposed to curved, blades. The cycloturbine variety has variable pitch to reduce the torque pulsation and is self-starting. The advantages of variable pitch are: high starting torque; a wide, relatively flat torque curve; a higher coefficient of performance; more efficient operation in turbulent winds; and a lower blade speed ratio which lowers blade bending stresses. Straight, V, or curved blades may be used.

Savonius Wind Turbine

These are drag-type devices with two (or more) scoops that are used in anemometers, *Flettner* vents (commonly seen on bus and van roofs), and in some high-reliability low-efficiency power turbines. They are always self-starting if there are at least three scoops.

Twisted Savonius

Twisted Savonius is a modified savonius, with long helical scoops to provide smooth torque. This is often used as a rooftop windturbine and has even been adapted for ships.

Another type of vertical axis is the Parallel turbine, which is similar to the crossflow fan or centrifugal fan. It uses the ground effect. Vertical axis turbines of this type have been tried for many years: a unit producing 10 kW was built by Israeli wind pioneer Bruce Brill in the 1980s.

Vortexis

The most recent advancement in Vertical Axis Wind Turbines has been the Vortexis VAWT, utilizing a pre-swirled augmented vertical axis wind turbine (PA-VAWT) designed for the purpose of developing a high efficiency VAWT concept that keeps the advantages of VAWT's compact size, lack of bias as to incoming wind direction, easy deployment and low radar cross section for use in mobile applications for the military, referred to in Special Operations as "Black Swan."

Design and Construction

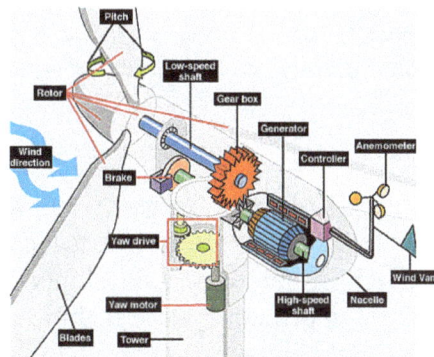

Components of a horizontal-axis wind turbine

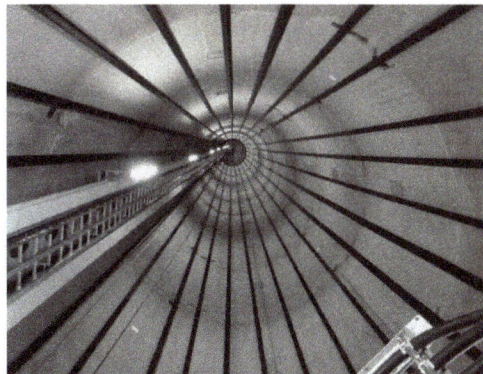

Inside view of a wind turbine tower, showing the tendon cables.

Wind turbines are designed to exploit the wind energy that exists at a location. Aerodynamic modeling is used to determine the optimum tower height, control systems, number of blades and blade shape.

Wind turbines convert wind energy to electricity for distribution. Conventional horizontal axis turbines can be divided into three components:

- The rotor component, which is approximately 20% of the wind turbine cost, includes the blades for converting wind energy to low speed rotational energy.

- The generator component, which is approximately 34% of the wind turbine cost, includes the electrical generator, the control electronics, and most likely a gearbox (e.g. planetary gearbox), adjustable-speed drive or continuously variable transmission component for converting the low speed incoming rotation to high speed rotation suitable for generating electricity.

- The structural support component, which is approximately 15% of the wind turbine cost, includes the tower and rotor yaw mechanism.

A 1.5 MW wind turbine of a type frequently seen in the United States has a tower 80 meters (260 ft) high. The rotor assembly (blades and hub) weighs 22,000 kilograms (48,000 lb). The nacelle, which contains the generator component, weighs 52,000 kilograms (115,000 lb). The concrete base for the tower is constructed using 26,000 kilograms (58,000 lb) of reinforcing steel and contains 190 cubic meters (250 cu yd) of concrete. The base is 15 meters (50 ft) in diameter and 2.4 meters (8 ft) thick near the center.

Among all renewable energy systems wind turbines have the highest effective intensity of power-harvesting surface because turbine blades not only harvest wind power, but also concentrate it.

Unconventional Designs

An E-66 wind turbine in the Windpark Holtriem, Germany, has an observation deck for visitors. Another turbine of the same type with an observation deck is located in Swaffham, England. Airborne wind turbine designs have been proposed and developed for many years but have yet to produce significant amounts of energy. In principle, wind turbines may also be used in conjunction with a large vertical solar updraft tower to extract the energy due to air heated by the sun.

The corkscrew shaped wind turbine at Progressive Field in Cleveland, Ohio

Wind turbines which utilise the Magnus effect have been developed.

A ram air turbine (RAT) is a special kind of small turbine that is fitted to some aircraft. When deployed, the RAT is spun by the airstream going past the aircraft and can provide power for the most essential systems if there is a loss of all on-board electrical power, as in the case of the "Gimli Glider".

The two-bladed turbine SCD 6MW offshore turbine designed by aerodyn Energiesysteme, built by MingYang Wind Power has a helideck for helicopters on top of its nacelle. The prototype was erected in 2014 in Rudong China.

Turbine Monitoring and Diagnostics

Due to data transmission problems, structural health monitoring of wind turbines is usually performed using several accelerometers and strain gages attached to the nacelle to monitor the gearbox and equipments. Currently, digital image correlation and stereophotogrammetry are used to measure dynamics of wind turbine blades. These methods usually measure displacement and strain to identify location of defects. Dynamic characteristics of non-rotating wind turbines have been measured using digital image correlation and photogrammetry. Three dimensional point tracking has also been used to measure rotating dynamics of wind turbines.

Materials and Durability

Currently serving wind turbine blades are mainly made of composite materials. These blades are usually made of a polyester resin, a vinyl resin, and epoxy thermosetting matrix resin and E-glass fibers, S- glass fibers and carbon fiber reinforced materials. Construction may use manual layup techniques or composite resin injection molding. As the price of glass fibers is only about one tenth the price of carbon fiber, glass fiber is still dominant. One of the predominant ways wind turbines have gain performance is by increasing rotor diameters, and thus blade length. Longer blades place more demands on the strength and stiffness of the materials. Stiffness is especially important to avoid having blades flex to the degree that they hit the tower of the wind turbine. Carbon fiber is between 4 and 6 times stiffer than glass fiber, so carbon fiber is becoming more common in wind turbine blades.

Wind Turbines on Public Display

The Nordex N50 wind turbine and visitor centre of Lamma Winds in Hong Kong, China

A few localities have exploited the attention-getting nature of wind turbines by placing them on public display, either with visitor centers around their bases, or with viewing areas farther away. The wind turbines are generally of conventional horizontal-axis, three-bladed design, and generate power to feed electrical grids, but they also serve the unconventional roles of technology demonstration, public relations, and education.

Small Wind Turbines

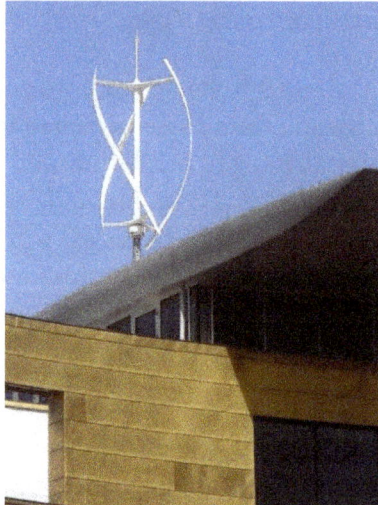

A small Quietrevolution QR5 Gorlov type vertical axis wind turbine in Bristol, England. Measuring 3 m in diameter and 5 m high, it has a nameplate rating of 6.5 kW to the grid.

Small wind turbines may be used for a variety of applications including on- or off-grid residences, telecom towers, offshore platforms, rural schools and clinics, remote monitoring and other purposes that require energy where there is no electric grid, or where the grid is unstable. Small wind turbines may be as small as a fifty-watt generator for boat or caravan use. Hybrid solar and wind powered units are increasingly being used for traffic signage, particularly in rural locations, as they avoid the need to lay long cables from the nearest mains connection point. The U.S. Department of Energy's National Renewable Energy Laboratory (NREL) defines small wind turbines as those smaller than or equal to 100 kilowatts. Small units often have direct drive generators, direct current output, aeroelastic blades, lifetime bearings and use a vane to point into the wind.

Larger, more costly turbines generally have geared power trains, alternating current output, flaps and are actively pointed into the wind. Direct drive generators and aeroelastic blades for large wind turbines are being researched.

Wind Turbine Spacing

On most horizontal windturbine farms, a spacing of about 6-10 times the rotor diameter is often upheld. However, for large wind farms distances of about 15 rotor diameters should be more economically optimal, taking into account typical wind turbine and land costs. This conclusion has been reached by research conducted by Charles Meneveau of the Johns Hopkins University, and Johan Meyers of Leuven University in Belgium, based on computer simulations that take into account the detailed interactions among wind turbines (wakes) as well as with the entire turbulent atmospheric boundary layer. Moreover, recent research by John Dabiri of Caltech suggests that vertical wind turbines may be placed much more closely together so long as an alternating pattern of rotation is created allowing blades of neighbouring turbines to move in the same direction as they approach one another.

Operability

Maintenance

Wind turbines need regular maintenance to stay reliable and available, reaching 98%.

Modern turbines usually have a small onboard crane for hoisting maintenance tools and minor components. However, large heavy components like generator, gearbox, blades and so on are rarely replaced and a heavy lift external crane is needed in those cases. If the turbine has a difficult access road, a containerized crane can be lifted up by the internal crane to provide heavier lifting.

Repowering

Installation of new wind turbines can be controversial. An alternative is repowering, where existing wind turbines are replaced with bigger, more powerful ones, sometimes in smaller numbers while keeping or increasing capacity.

Demolition

Older turbines were in some early cases not required to be removed when reaching the end of their life. Some still stand, waiting to be recycled or repowered.

A demolition industry develops to recycle offshore turbines at a cost of DKK 2-4 million per MW, to be guaranteed by the owner.

Records

Fuhrländer Wind Turbine Laasow, in Brandenburg, Germany, among the world's tallest wind turbines

Éole, the largest vertical axis wind turbine, in Cap-Chat, Quebec, Canada

Largest capacity conventional drive

The Vestas V164 has a rated capacity of 8.0 MW, has an overall height of 220 m (722 ft), a diameter of 164 m (538 ft), is for offshore use, and is the world's largest-capacity wind turbine since its introduction in 2014. The conventional drive train consist of a main gearbox and a medium speed PM generator. Prototype installed in 2014 at the National Test Center Denmark nearby Østerild. Series production starts end of 2015.

Largest capacity direct drive

The Enercon E-126 with 7.58MW and 127m rotor diameter is the largest direct drive turbine. It's only for onshore use. The turbine has parted rotor blades with 2 sections for transport. In July 2016, Siemens upgraded its 7MW to 8MW.

Largest vertical-axis

Le Nordais wind farm in Cap-Chat, Quebec has a vertical axis wind turbine (VAWT) named Éole, which is the world's largest at 110 m. It has a nameplate capacity of 3.8 MW.

Largest 1 bladed turbine

Riva Calzoni M33 was a Single Bladed Wind Turbine with 350 kW, designed and built In Bologna in 1993

Largest 2 bladed turbine

Today's biggest 2 bladed turbine is build by Mingyang Wind Power in 2013. It is a SCD6.5MW offshore downwind turbine, designed by aerodyn Energiesysteme

Largest swept area

The turbine with the largest swept area is the Samsung S7.0-171, with a diameter of 171 m, giving a total sweep of 22966 m^2.

Tallest

A Nordex 3.3 MW was installed in July 2016. It has a total height of 230m, and a hub height of 164m on 100m concrete tower bottom with steel tubes on top (hybrid tower).

Vestas V164 was the tallest wind turbine, standing in Østerild, Denmark, 220 meters tall, constructed in 2014. It has a steel tube tower.

Highest tower

Fuhrländer installed a 2.5MW turbine on a 160m lattice tower in 2003.

Most rotors

Lagerwey has build Four-in-One, a multi rotor wind turbine with one tower and four rotors near Maasvlakte. In April 2016, Vestas installed a 900kW quadrotor test wind turbine at Risø, made from 4 recycled 225kW V29 turbines.

Most productive

Four turbines at Rønland wind farm in Denmark share the record for the most productive wind turbines, with each having generated 63.2 GWh by June 2010.

Highest-situated

Since 2013 the world's highest-situated wind turbine was made and installed by WindAid and is located at the base of the Pastoruri Glacier in Peru at 4,877 meters (16,001 ft) above sea level. The site uses the WindAid 2.5 kW wind generator to supply power to a small rural community of micro entrepreneurs who cater to the tourists who come to the Pastoruri glacier.

Largest floating wind turbine

The world's largest—and also the first operational deep-water *large-capacity*—floating wind turbine is the 2.3 MW Hywind currently operating 10 kilometers (6.2 mi) offshore in 220-meter-deep water, southwest of Karmøy, Norway. The turbine began operating in September 2009 and utilizes a Siemens 2.3 MW turbine.

Modification in Wind Turbine

Wind Lens

A 3kW wind-lens turbine near Fukuoka Bay in 2012

A wind lens is a modification made to a wind turbine to make it a more efficient way to capture wind energy. The modification is a ring structure called a "brim" or "wind lens" which surrounds the blades, diverting air away from the exhaust outflow behind the blades. The turbulence created as a result of the new configuration creates a low pressure zone behind the turbine, causing greater wind to pass through the turbine, and this, in turn, increases blade rotation and energy output. In Japan, wind lenses are being researched by the Wind Engineering Section of Kyushu University.

Their website claims:

Wind power is proportional to the wind speed cubed. If we can increase the wind speed with some mechanism by utilizing the fluid dynamic nature around a structure, namely if we can capture and concentrate the wind energy locally, the output power of a wind turbine can be increased substantially. At wind energy section of Kyushu University, a new efficient wind power turbine system has been developed. This system has a diffuser shroud at the circumference of its rotor to embody the wind energy concentration. The diffuser shroud is now named "Wind lens". To apply the wind-lens structure to a larger size turbine, we have developed a compact collection-acceleration device. There are several ongoing projects in which the wind-lens turbines are involved.

Test trials are under way. One report suggests the modification results in twice or three times as much power as conventional wind turbines.

Compact Wind Acceleration Turbine

Compact Wind Acceleration Turbines (CWATs) are a class of wind turbine that uses structures to accelerate wind before it enters the wind-generating element. The concept of these structures has been around for decades but has not gained wide acceptance in the marketplace. In 2008, two companies targeting the mid-wind (100 kW-1 MW) marketplace have received funding from venture capital. The first company to receive funding is Optiwind, which received its series A funding in April 2008, and the second company is Ogin, Inc. (formerly FloDesign Wind Turbine Inc.), which also received its series A funding in April 2008. Optiwind is funded through Charles River Ventures and FloDesign is funded through Kleiner Perkins. Other CWATs under development include the WindTamer from AristaPower, WindCube, Innowind (conceptual offshore application) and Enflo turbines.

History

CWATs are a new acronym that encompasses the class of machines formerly known as DAWTs (diffuser augmented wind turbines). The technologies mentioned above all use diffuser augmentation that is substantially similar to previous designs as the primary means of acceleration. DAWTs were heavily researched by K. Foreman and Oman of Grumman Aerospace in the 1970s and 1980s and Igra in Israel in the 1970s. At the end of a decade of wind tunnel research and design funded by Grumman, NASA, and the DOE, it was determined that the DAWT system's economics were not sufficient to justify commercialization. In the 1990s the Grumman technology was licensed to a New Zealand company, Vortec Wind. The attempt to commercialize the Vortec 7 in New Zealand from 1998 to 2002 proved it to be economically untenable when compared to the dominant HAWT (horizontal axis wind turbine) technology.

Economics

Ultimately, any wind turbine design must be measured against economic realities. It must positively answer the question, "is the cost to install and operate the system lower than the cost of other alternatives, including the local electric grid?" Historically, CWAT/DAWT designs have failed to overcome this hurdle as compared to more conventional HAWT designs. However, there is reason to believe that this equation may be shifting towards these new designs. The two primary drivers of this equation have been the amount of augmentation and the structural implications of these additional design elements.

Augmentation

The first factor regards power increase and the method of comparison used by DAWT (and more recently CWAT) designers to determine whether the system is worth developing. Grumman and other attempts to commercialize these machines compare their machines to HAWTs based on a rotor area to rotor area comparison. As Van Bussel of Delft (The Science of Making More Torque from Wind: Diffuser Experiments and Theory Revisited, G.J.W. van Bussel, Delft, 2007) pointed out, this is an inaccurate comparison and the comparison of power multiples should be made on the basis of the exit area of the diffuser or shroud not the rotor area. Grumman claimed a 4× increase over an unshrouded turbine based on an acceleration of 1.6 times the ambient wind velocity (An Investigation on Diffuser Augmented Turbines, D.G. Philips, 2003 (reference materials compiled from K.M. Foreman)). A 1.6 acceleration is in fact 2.6 times the power of a HAWT if the ratio of the shrouded rotor to the exit area is 1.6. If however the rotor to exit area ratio is 2.75 (as it was in the Grumman case), the actual power increase over a HAWT with the same swept area as the diffuser exit area is only 1.4× the power (a Cp of .34 related to the exit area, slightly better than a small unducted wind turbine and significantly worse than utility scale wind turbine Cp's). Given that the DAWTs with this ratio have roughly a solidity of 60+% when the blades and the diffuser are accounted for and the solidity of the HAWT is roughly 10%, the cost and amount of material needed to produce the 40% gain outweighs the increase in power.

Structural Implications

Second is the structural requirement in terms of resisting overturning and bending in extreme wind events which all wind turbines must be designed for in accordance with the IEC 61400 series of standards (IEC). The DAWT structure typically has poor drag characteristics. That combined with higher solidity can lead to substantially greater structural costs than a HAWT in the support structure, the yaw bearing, and the foundation, when using conventional monopole designs. However, the advent of new tower designs, flange geometries and foundation systems appear to be successfully challenging these historical norms but if so then these advances can be equally well applied to improving the economics of conventional HAWT designs.

Optiwind

In the case of Optiwind (now defunct), there appears to be a growing body of evidence that they believe have solved for both the acceleration and economic challenges posed by CWAT/DAWT designs. Where previous attempts at new designs in this category have focused purely on acceleration magnitude, Optiwind appears to have taken a more holistic approach to their design,

considering cost as much as acceleration benefit. In addition, the ongoing operational and maintenance cost of the entire unit appears to be successfully addressed in this design. It is absent the significant cost drivers of HAWT systems - large composite blades, gearbox, yaw motor, pitch control,
lubrication, etc. In addition, the novel foundation geometry appears to have mitigated the structural
challenges of the conventional monopole foundation design, which was originally conceived to offset
the counter-rotational effects ("wobbling") of large, three blade turbines. As such, it is reasonable
to assume that Optiwind's holistic approach to systemwide costs have led to a series of designs and
discoveries that can realistically deliver the economic advantages of accelerated wind at a cost that is
less than the net system cost. This is accurate if the Optiwind system is compared to a HAWT purely
on rating. The problem therein is, if one compares the Optiwind design based on its stack height (the
distance from its lowest turbine to the highest turbine) to a traditional HAWT of the same diameter,
the overall power output of the machine is 20% less than that of the HAWT, with all the attendant
material and structural expenses generally associated with a CWAT/DAWT. On average a CWAT/
DAWT system would need to produce at least 2-3 times the energy a HAWT could produce from the
maximum area used by the CWAT/DAWT in order to offset the substantially larger material costs.
There is no evidence yet that there are any DAWT/CWAT designs capable of this level of increase
when compared on an apples to apples basis with HAWT's.

Ogin (Formerly Flodesign Wind Turbine)

Ogin's MEWT (mixer-ejector wind turbine, another CWAT variation) is differentiated from previous
DAWT's by using a lobed two stage diffuser (Grumman and Vortec machine were also two stage, but
conical instead of lobed) to equalize the pressure over the exit area of the diffuser. The theory is that
creating a uniform pressure distribution with the lobes and the injection of external flow will prevent
boundary layer separation in the diffuser thereby allowing the maximum acceleration through rotor.
Werle and Presz's (Flodesign's chief scientists) paper, AAIA technical note Ducted Water/Wind turbines revisited - 2007, details the theory behind their design. Maximum acceleration detailed in their
paper is 1.8× the ambient velocity from which they derive that 3 times more power is available at the
rotor than for an unshrouded turbine. When referred to exit area this multiple drops to parity with the
HAWT power. Ogin's turbine based on released images and CAD's appears to be substantially similar
to the Vortec and Grumman machines except for the lobed inner annulus. This would indicate that its
drag characteristics can be expected to be similar. Newer information on the Ogin website (www.ogine-
nergy.com) shows the lobes flattened out into 2D panels.

Performance

The science of wind acceleration around a structure, as well as the vortex shedding benefits of a
shroud/diffuser, are well understood and tested. From Bernoulli forward, science has substantially vetted these concepts and there is general academic consensus as to their veracity and their
potential impact on wind power production. DAWT's however have the classic boundary layer
separation problem experienced by airfoils at a "stall" angle of attack. This significantly reduces
the acceleration achievable by a DAWT relative to the theoretical rate indicated by its exit to area
ratio, per Flodesign paper mentioned above. It is generally thought that since the amount of power
produced by a wind turbine is proportional to the cube of the wind speed, any acceleration benefit
is potentially statistically significant in the economics of wind. As noted though this is an inaccurate as it ignores the impact of the exit to area ratio and is therefore an apples to oranges compar-

ison. In the case of a typical CWAT/DAWT the power result in perfect theoretical operation once adjusted for the area of the shroud is actually the square of the velocity at the rotor. As the CWAT/DAWT diverges from theoretical function the power increase drops significantly according to the formula derived from mass conservation,

Power ratio DAWT to HAWT = (Athroat/Aintake)(vthroat/vfreestream)³

Power ratio DAWT to HAWT = (1/2.75)(27.5 ms/10 ms)³ = 7.56 increase

So for example, a DAWT operating at theoretical function of 1.8 with a 2.75 area ratio per Flodesign,

Power ratio DAWT to HAWT = (1/2.75)(18 ms/10 ms)³ = 2.12 increase

For the highest claimed velocity increase in a DAWT of 1.6 × freestream,

Power ratio DAWT to HAWT = (1/2.75)(16 ms/10 ms)³ = 1.48 increase

Not significant enough to offset the associated costs. The problem with optiwind is even more severe since the system only covers a fraction of the swept area available to a HAWT of the stack height.

The challenge has always been, and remains, installing, operating, and maintaining these structures for a cost that is less than the incremental value gained by their presence. Recent developments in material science, installation methodology and overall system integration have led to the far more realistic view that we are very close to this advent and the dawn of a new, highly sustainable class of wind turbine if the issues elucidated above can be dealt with which still remains highly questionable for the DAWT geometry.

Among the recent DAWT designs that appear to have a definitive positive power, if not cost, comparison to HAWTs is the Enflo turbine. Based on its rotor:exit ratio and the published power performance this turbine appears to have a confirmed 2 times increase in power output over a HAWT of the diameter of the exit area. It is still unlikely that this machine can scale to larger ratings but based on their published data (not confirmed by third party testing) the Enflo appears to be the best performing DAWT/CWAT yet built.

Hydroelectricity

The Three Gorges Dam in Central China is the world's largest power producing facility of any kind.

Hydroelectricity is the term referring to electricity generated by hydropower; the production of electrical power through the use of the gravitational force of falling or flowing water. In 2015 hydropower generated 16.6% of the worlds total electricity and 70% of all renewable electricity, and is expected to increase about 3.1% each year for the next 25 years.

Hydropower is produced in 150 countries, with the Asia-Pacific region generating 33 percent of global hydropower in 2013. China is the largest hydroelectricity producer, with 920 TWh of production in 2013, representing 16.9 percent of domestic electricity use.

The cost of hydroelectricity is relatively low, making it a competitive source of renewable electricity. The hydro station consumes no water, unlike coal or gas plants. The average cost of electricity from a hydro station larger than 10 megawatts is 3 to 5 U.S. cents per kilowatt-hour. With a dam and reservoir it is also a flexible source of electricity since the amount produced by the station can be changed up or down very quickly to adapt to changing energy demands. Once a hydroelectric complex is constructed, the project produces no direct waste, and has a considerably lower output level of greenhouse gases than fossil fuel powered energy plants.

History

Museum Hydroelectric power plant "Under the Town" in Serbia, built in 1900.

Hydropower has been used since ancient times to grind flour and perform other tasks. In the mid-1770s, French engineer Bernard Forest de Bélidor published *Architecture Hydraulique* which described vertical- and horizontal-axis hydraulic machines. By the late 19th century, the electrical generator was developed and could now be coupled with hydraulics. The growing demand for the Industrial Revolution would drive development as well. In 1878 the world's first hydroelectric power scheme was developed at Cragside in Northumberland, England by William George Armstrong. It was used to power a single arc lamp in his art gallery. The old Schoelkopf Power Station No. 1 near Niagara Falls in the U.S. side began to produce electricity in 1881. The first Edison hydroelectric power station, the Vulcan Street Plant, began operating September 30, 1882, in Appleton, Wisconsin, with an output of about 12.5 kilowatts. By 1886 there were 45 hydroelectric power stations in the U.S. and Canada. By 1889 there were 200 in the U.S. alone.

At the beginning of the 20th century, many small hydroelectric power stations were being constructed by commercial companies in mountains near metropolitan areas. Grenoble, France held the International Exhibition of Hydropower and Tourism with over one million visitors. By 1920 as 40% of the power produced in the United States was hydroelectric, the Federal Power Act was enacted into law. The Act created the Federal Power Commission to regulate hydroelectric power stations on federal land and water. As the power stations became larger, their associated dams developed additional purposes to include flood control, irrigation and navigation. Federal funding became necessary for large-scale development and federally owned corporations, such as the Tennessee Valley Authority (1933) and the Bonneville Power Administration (1937) were created. Additionally, the Bureau of Reclamation which had begun a series of western U.S. irrigation projects in the early 20th century was now constructing large hydroelectric projects such as the 1928 Hoover Dam. The U.S. Army Corps of Engineers was also involved in hydroelectric development, completing the Bonneville Dam in 1937 and being recognized by the Flood Control Act of 1936 as the premier federal flood control agency.

Hydroelectric power stations continued to become larger throughout the 20th century. Hydropower was referred to as *white coal* for its power and plenty. Hoover Dam's initial 1,345 MW power station was the world's largest hydroelectric power station in 1936; it was eclipsed by the 6809 MW Grand Coulee Dam in 1942. The Itaipu Dam opened in 1984 in South America as the largest, producing 14,000 MW but was surpassed in 2008 by the Three Gorges Dam in China at 22,500 MW. Hydroelectricity would eventually supply some countries, including Norway, Democratic Republic of the Congo, Paraguay and Brazil, with over 85% of their electricity. The United States currently has over 2,000 hydroelectric power stations that supply 6.4% of its total electrical production output, which is 49% of its renewable electricity.

Generating Methods

Turbine row at El Nihuil II Power Station in Mendoza, Argentina

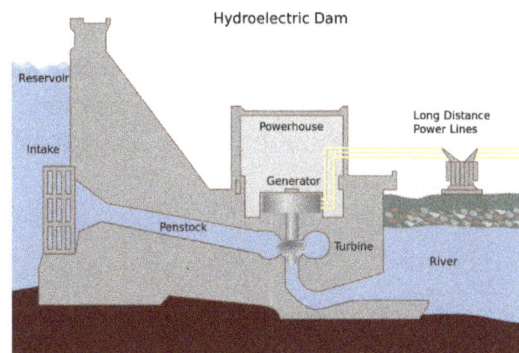

Cross section of a conventional hydroelectric dam.

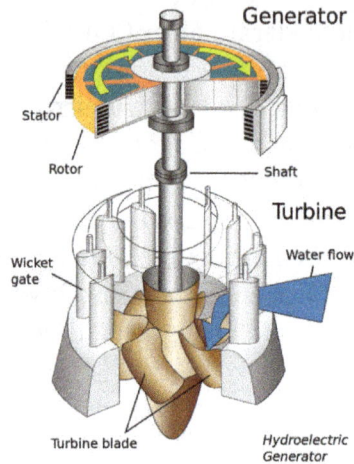

A typical turbine and generator

Conventional (Dams)

Most hydroelectric power comes from the potential energy of dammed water driving a water turbine and generator. The power extracted from the water depends on the volume and on the difference in height between the source and the water's outflow. This height difference is called the head. A large pipe (the "penstock") delivers water from the reservoir to the turbine.

Pumped-Storage

This method produces electricity to supply high peak demands by moving water between reservoirs at different elevations. At times of low electrical demand, the excess generation capacity is used to pump water into the higher reservoir. When the demand becomes greater, water is released back into the lower reservoir through a turbine. Pumped-storage schemes currently provide the most commercially important means of large-scale grid energy storage and improve the daily capacity factor of the generation system. Pumped storage is not an energy source, and appears as a negative number in listings.

Run-of-the-River

Run-of-the-river hydroelectric stations are those with small or no reservoir capacity, so that only the water coming from upstream is available for generation at that moment, and any oversupply must pass unused. A constant supply of water from a lake or existing reservoir upstream is a significant advantage in choosing sites for run-of-the-river. In the United States, run of the river hydropower could potentially provide 60,000 megawatts (80,000,000 hp) (about 13.7% of total use in 2011 if continuously available).

Tide

A tidal power station makes use of the daily rise and fall of ocean water due to tides; such sources are highly predictable, and if conditions permit construction of reservoirs, can also be dispatchable to generate power during high demand periods. Less common types of hydro schemes use water's kinetic energy or undammed sources such as undershot water wheels. Tidal power is viable in a relatively small number of locations around the world. In Great Britain, there are eight sites that could be developed, which have the potential to generate 20% of the electricity used in 2012.

Sizes, Types and Capacities of Hydroelectric Facilities

Large Facilities

Large-scale hydroelectric power stations are more commonly seen as the largest power producing facilities in the world, with some hydroelectric facilities capable of generating more than double the installed capacities of the current largest nuclear power stations.

Although no official definition exists for the capacity range of large hydroelectric power stations, facilities from over a few hundred megawatts are generally considered large hydroelectric facilities.

Currently, only four facilities over 10 GW (10,000 MW) are in operation worldwide, see table below.

Rank	Station	Country	Location	Capacity (MW)
1.	Three Gorges Dam	China	30°49′15″N 111°00′08″E30.82083°N 111.00222°E	22,500
2.	Itaipu Dam	Brazil Paraguay	25°24′31″S 54°35′21″W25.40861°S 54.58917°W	14,000
3.	Xiluodu Dam	China	28°15′35″N 103°38′58″E28.25972°N 103.64944°E	13,860
4.	Guri Dam	Venezuela	07°45′59″N 62°59′57″W7.76639°N 62.99917°W	10,200

Panoramic view of the Itaipu Dam, with the spillways (closed at the time of the photo) on the left. In 1994, the American Society of Civil Engineers elected the Itaipu Dam as one of the seven modern Wonders of the World.

Small

Small hydro is the development of hydroelectric power on a scale serving a small community or industrial plant. The definition of a small hydro project varies but a generating capacity of up to 10 megawatts (MW) is generally accepted as the upper limit of what can be termed small hydro. This may be stretched to 25 MW and 30 MW in Canada and the United States. Small-scale hydroelectricity production grew by 28% during 2008 from 2005, raising the total world small-hydro capacity to 85 GW. Over 70% of this was in China (65 GW), followed by Japan (3.5 GW), the United States (3 GW), and India (2 GW).

A micro-hydro facility in Vietnam

Pico hydroelectricity in Mondulkiri, Cambodia

Small hydro stations may be connected to conventional electrical distribution networks as a source of low-cost renewable energy. Alternatively, small hydro projects may be built in isolated areas that would be uneconomic to serve from a network, or in areas where there is no national electrical distribution network. Since small hydro projects usually have minimal reservoirs and civil construction work, they are seen as having a relatively low environmental impact compared to large hydro. This decreased environmental impact depends strongly on the balance between stream flow and power production.

Micro

Micro hydro is a term used for hydroelectric power installations that typically produce up to 100 kW of power. These installations can provide power to an isolated home or small community, or are sometimes connected to electric power networks. There are many of these installations around the world, particularly in developing nations as they can provide an economical source of energy without purchase of fuel. Micro hydro systems complement photovoltaic solar energy systems because in many areas, water flow, and thus available hydro power, is highest in the winter when solar energy is at a minimum.

Pico

Pico hydro is a term used for hydroelectric power generation of under 5 kW. It is useful in small, remote communities that require only a small amount of electricity. For example, to power one or two fluorescent light bulbs and a TV or radio for a few homes. Even smaller turbines of 200-300W may power a single home in a developing country with a drop of only 1 m (3 ft). A Pico-hydro setup is typically run-of-the-river, meaning that dams are not used, but rather pipes divert some of the flow, drop this down a gradient, and through the turbine before returning it to the stream.

Underground

An underground power station is generally used at large facilities and makes use of a large natural height difference between two waterways, such as a waterfall or mountain lake. An underground tunnel is constructed to take water from the high reservoir to the generating hall built in an underground cavern near the lowest point of the water tunnel and a horizontal tailrace taking water away to the lower outlet waterway.

Measurement of the tailrace and forebay rates at the Limestone Generating Station in Manitoba, Canada.

Calculating Available Power

A simple formula for approximating electric power production at a hydroelectric station is: $P = \rho hrgk$, where

- P is Power in watts,

- ρ is the density of water (~1000 kg/m³),

- h is height in meters,

- r is flow rate in cubic meters per second,

- g is acceleration due to gravity of 9.8 m/s²,

- k is a coefficient of efficiency ranging from 0 to 1. Efficiency is often higher (that is, closer to 1) with larger and more modern turbines.

Annual electric energy production depends on the available water supply. In some installations, the water flow rate can vary by a factor of 10:1 over the course of a year.

Advantages and Disadvantages

Advantages

The Ffestiniog Power Station can generate 360 MW of electricity within 60 seconds of the demand arising.

Flexibility

Hydropower is a flexible source of electricity since stations can be ramped up and down very quickly to adapt to changing energy demands. Hydro turbines have a start-up time of the order of a few minutes. It takes around 60 to 90 seconds to bring a unit from cold start-up to full load; this is much shorter than for gas turbines or steam plants. Power generation can also be decreased quickly when there is a surplus power generation. Hence the limited capacity of hydropower units is not generally used to produce base power except for vacating the flood pool or meeting downstream needs. Instead, it serves as backup for non-hydro generators.

Low Power Costs

The major advantage of hydroelectricity is elimination of the cost of fuel. The cost of operating a hydroelectric station is nearly immune to increases in the cost of fossil fuels such as oil, natural gas or coal, and no imports are needed. The average cost of electricity from a hydro station larger than 10 megawatts is 3 to 5 U.S. cents per kilowatt-hour.

Hydroelectric stations have long economic lives, with some plants still in service after 50–100 years. Operating labor cost is also usually low, as plants are automated and have few personnel on site during normal operation.

Where a dam serves multiple purposes, a hydroelectric station may be added with relatively low construction cost, providing a useful revenue stream to offset the costs of dam operation. It has been calculated that the sale of electricity from the Three Gorges Dam will cover the construction costs after 5 to 8 years of full generation. Additionally, some data shows that in most countries large hydropower dams will be too costly and take too long to build to deliver a positive risk adjusted return, unless appropriate risk management measures are put in place.

Suitability for Industrial Applications

While many hydroelectric projects supply public electricity networks, some are created to serve specific industrial enterprises. Dedicated hydroelectric projects are often built to provide the substantial amounts of electricity needed for aluminium electrolytic plants, for example. The Grand Coulee Dam switched to support Alcoa aluminium in Bellingham, Washington, United States for

American World War II airplanes before it was allowed to provide irrigation and power to citizens (in addition to aluminium power) after the war. In Suriname, the Brokopondo Reservoir was constructed to provide electricity for the Alcoa aluminium industry. New Zealand's Manapouri Power Station was constructed to supply electricity to the aluminium smelter at Tiwai Point.

Reduced CO_2 Emissions

Since hydroelectric dams do not burn fossil fuels, they do not directly produce carbon dioxide. While some carbon dioxide is produced during manufacture and construction of the project, this is a tiny fraction of the operating emissions of equivalent fossil-fuel electricity generation. One measurement of greenhouse gas related and other externality comparison between energy sources can be found in the ExternE project by the Paul Scherrer Institute and the University of Stuttgart which was funded by the European Commission. According to that study, hydroelectricity produces the least amount of greenhouse gases and externality of any energy source. Coming in second place was wind, third was nuclear energy, and fourth was solar photovoltaic. The low greenhouse gas impact of hydroelectricity is found especially in temperate climates. The above study was for local energy in Europe; presumably similar conditions prevail in North America and Northern Asia, which all see a regular, natural freeze/thaw cycle (with associated seasonal plant decay and regrowth). Greater greenhouse gas emission impacts are found in the tropical regions because the reservoirs of power stations in tropical regions produce a larger amount of methane than those in temperate areas.

Other Uses of The Reservoir

Reservoirs created by hydroelectric schemes often provide facilities for water sports, and become tourist attractions themselves. In some countries, aquaculture in reservoirs is common. Multi-use dams installed for irrigation support agriculture with a relatively constant water supply. Large hydro dams can control floods, which would otherwise affect people living downstream of the project.

Disadvantages

Ecosystem damage and loss of land

Hydroelectric power stations that use dams would submerge large areas of land due to the requirement of a reservoir. Merowe Dam in Sudan.

Large reservoirs associated with traditional hydroelectric power stations result in submersion of extensive areas upstream of the dams, sometimes destroying biologically rich and productive lowland and riverine valley forests, marshland and grasslands. Damming interrupts the flow of rivers and can harm local ecosystems, and building large dams and reservoirs often involves displacing people and wildlife. The loss of land is often exacerbated by habitat fragmentation of surrounding areas caused by the reservoir.

Hydroelectric projects can be disruptive to surrounding aquatic ecosystems both upstream and downstream of the plant site. Generation of hydroelectric power changes the downstream river environment. Water exiting a turbine usually contains very little suspended sediment, which can lead to scouring of river beds and loss of riverbanks. Since turbine gates are often opened intermittently, rapid or even daily fluctuations in river flow are observed.

Siltation and Flow Shortage

When water flows it has the ability to transport particles heavier than itself downstream. This has a negative effect on dams and subsequently their power stations, particularly those on rivers or within catchment areas with high siltation. Siltation can fill a reservoir and reduce its capacity to control floods along with causing additional horizontal pressure on the upstream portion of the dam. Eventually, some reservoirs can become full of sediment and useless or over-top during a flood and fail.

Changes in the amount of river flow will correlate with the amount of energy produced by a dam. Lower river flows will reduce the amount of live storage in a reservoir therefore reducing the amount of water that can be used for hydroelectricity. The result of diminished river flow can be power shortages in areas that depend heavily on hydroelectric power. The risk of flow shortage may increase as a result of climate change. One study from the Colorado River in the United States suggest that modest climate changes, such as an increase in temperature in 2 degree Celsius resulting in a 10% decline in precipitation, might reduce river run-off by up to 40%. Brazil in particular is vulnerable due to its heavy reliance on hydroelectricity, as increasing temperatures, lower water flow and alterations in the rainfall regime, could reduce total energy production by 7% annually by the end of the century.

Methane Emissions (From Reservoirs)

The Hoover Dam in the United States is a large conventional dammed-hydro facility, with an installed capacity of 2,080 MW.

Lower positive impacts are found in the tropical regions, as it has been noted that the reservoirs of power plants in tropical regions produce substantial amounts of methane. This is due to plant material in flooded areas decaying in an anaerobic environment, and forming methane, a greenhouse gas. According to the World Commission on Dams report, where the reservoir is large compared to the generating capacity (less than 100 watts per square metre of surface area) and no clearing of the forests in the area was undertaken prior to impoundment of the reservoir, greenhouse gas emissions from the reservoir may be higher than those of a conventional oil-fired thermal generation plant.

In boreal reservoirs of Canada and Northern Europe, however, greenhouse gas emissions are typically only 2% to 8% of any kind of conventional fossil-fuel thermal generation. A new class of underwater logging operation that targets drowned forests can mitigate the effect of forest decay.

Relocation

Another disadvantage of hydroelectric dams is the need to relocate the people living where the reservoirs are planned. In 2000, the World Commission on Dams estimated that dams had physically displaced 40-80 million people worldwide.

Failure Risks

Because large conventional dammed-hydro facilities hold back large volumes of water, a failure due to poor construction, natural disasters or sabotage can be catastrophic to downriver settlements and infrastructure. Dam failures have been some of the largest man-made disasters in history.

During Typhoon Nina in 1975 Banqiao Dam failed in Southern China when more than a year's worth of rain fell within 24 hours. The resulting flood resulted in the deaths of 26,000 people, and another 145,000 from epidemics. Millions were left homeless. Also, the creation of a dam in a geologically inappropriate location may cause disasters such as 1963 disaster at Vajont Dam in Italy, where almost 2,000 people died.

The Malpasset Dam failure in Fréjus on the French Riviera (Côte d'Azur), southern France, collapsed on December 2, 1959, killing 423 people in the resulting flood.

Smaller dams and micro hydro facilities create less risk, but can form continuing hazards even after being decommissioned. For example, the small Kelly Barnes Dam failed in 1967, causing 39 deaths with the Toccoa Flood, ten years after its power station was decommissioned the earthen embankment dam failed.

Comparison with Other Methods of Power Generation

Hydroelectricity eliminates the flue gas emissions from fossil fuel combustion, including pollutants such as sulfur dioxide, nitric oxide, carbon monoxide, dust, and mercury in the coal. Hydroelectricity also avoids the hazards of coal mining and the indirect health effects of coal emissions. Compared to nuclear power, hydroelectricity construction requires altering large areas of the environment while a nuclear power station has a small footprint, and hydro-powerstation failures have caused tens of thousands of more deaths than any nuclear station failure. The creation of Garrison Dam, for example, required Native American land to create Lake Sakakawea, which has a shoreline of 1,320 miles, and caused the inhabitants to sell 94% of their arable land for $7.5 million in 1949.

Compared to wind farms, hydroelectricity power stations have a more predictable load factor. If the project has a storage reservoir, it can generate power when needed. Hydroelectric stations can be easily regulated to follow variations in power demand.

World Hydroelectric Capacity

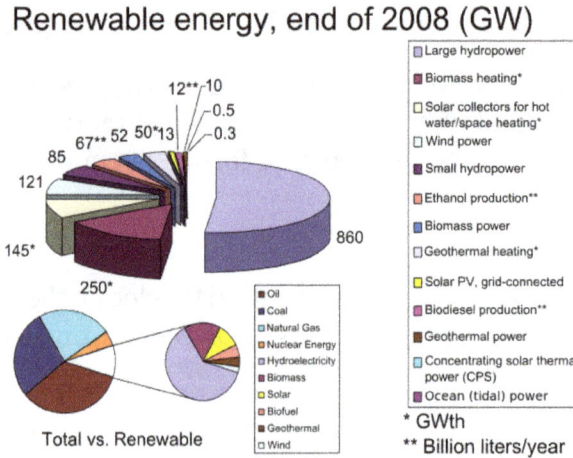

World renewable energy share (2008)

The ranking of hydro-electric capacity is either by actual annual energy production or by installed capacity power rating. In 2015 hydropower generated 16.6% of the worlds total electricity and 70% of all renewable electricity. Hydropower is produced in 150 countries, with the Asia-Pacific region generated 32 percent of global hydropower in 2010. China is the largest hydroelectricity producer, with 721 terawatt-hours of production in 2010, representing around 17 percent of domestic electricity use. Brazil, Canada, New Zealand, Norway, Paraguay, Austria, Switzerland, and Venezuela have a majority of the internal electric energy production from hydroelectric power. Paraguay produces 100% of its electricity from hydroelectric dams, and exports 90% of its production to Brazil and to Argentina. Norway produces 98–99% of its electricity from hydroelectric sources.

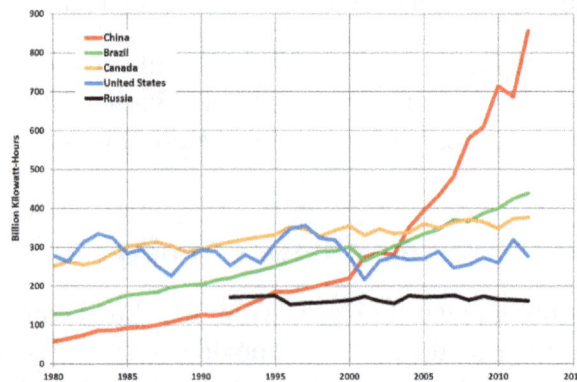

Trends in the top five hydroelectricity-producing countries

A hydro-electric station rarely operates at its full power rating over a full year; the ratio between annual average power and installed capacity rating is the capacity factor. The installed capacity is the sum of all generator nameplate power ratings.

Ten of the largest hydroelectric producers as at 2013.				
Country	Annual hydroelectric production (TWh)	Installed capacity (GW)	Capacity factor	% of total production
China	920	194	0.37	16.9%
Canada	392	76	0.59	60.1%
Brazil	391	86	0.56	68.6%
United States	290	102	0.42	6.7
Russia	183	50	0.42	17.3%
India	142	40	0.43	11.9%
Norway	129	31	0.49	96.1%
Japan	85	49	0.37	8.1%
Venezuela	84	15	0.67	67.8%
France	76	25	0.46	13.2%

Major Projects Under Construction

Name	Maximum Capacity	Country	Construction started	Scheduled completion	Comments
Belo Monte Dam	11,181 MW	Brazil	March, 2011	2015	Preliminary construction underway. Construction suspended 14 days by court order Aug 2012
Siang Upper HE Project	11,000 MW	India	April, 2009	2024	Multi-phase construction over a period of 15 years. Construction was delayed due to dispute with China.
Tasang Dam	7,110 MW	Burma	March, 2007	2022	Controversial 228 meter tall dam with capacity to produce 35,446 GWh annually.
Xiangjiaba Dam	6,400 MW	China	November 26, 2006	2015	The last generator was commissioned on July 9, 2014
Grand Ethiopian Renaissance Dam	6,000 MW	Ethiopia	2011	2017	Located in the upper Nile Basin, drawing complaint from Egypt
Nuozhadu Dam	5,850 MW	China	2006	2017	
Jinping 2 Hydropower Station	4,800 MW	China	January 30, 2007	2014	To build this dam, 23 families and 129 local residents need to be moved. It works with Jinping 1 Hydropower Station as a group.
Diamer-Bhasha Dam	4,500 MW	Pakistan	October 18, 2011	2023	

Name	Maximum Capacity	Country	Construction started	Scheduled completion	Comments
Jinping 1 Hydropower Station	3,600 MW	China	November 11, 2005	2014	The sixth and final generator was commissioned on 15 July 2014
Jirau Power Station	3,300 MW	Brazil	2008	2013	Construction halted in March 2011 due to worker riots.
Guanyinyan Dam	3,000 MW	China	2008	2015	Construction of the roads and spillway started.
Lianghekou Dam	3,000 MW	China	2014	2023	
Dagangshan Dam	2,600 MW	China	August 15, 2008	2016	
Liyuan Dam	2,400 MW	China	2008	2013	
Tocoma Dam Bolívar State	2,160 MW	Venezuela	2004	2014	This power station would be the last development in the Low Caroni Basin, bringing the total to six power stations on the same river, including the 10,000MW Guri Dam.
Ludila Dam	2,100 MW	China	2007	2015	Brief construction halt in 2009 for environmental assessment.
Shuangjiangkou Dam	2,000 MW	China	December, 2007	2018	The dam will be 312 m high.
Ahai Dam	2,000 MW	China	July 27, 2006	2015	
Teles Pires Dam	1,820 MW	Brazil	2011	2015	
Site C Dam	1,100 MW	Canada	2015	2024	First large dam in western Canada since 1984
Lower Subansiri Dam	2,000 MW	India	2007	2016	

Induction Generator

An induction generator or *asynchronous generator* is a type of alternating current (AC) electrical generator that uses the principles of induction motors to produce power. Induction generators operate by mechanically turning their rotors faster than synchronous speed. A regular AC asynchronous motor usually can be used as a generator, without any internal modifications. Induction generators are useful in applications such as mini hydro power plants, wind turbines, or in reducing high-pressure gas streams to lower pressure, because they can recover energy with relatively simple controls.

An induction generator usually draws its excitation power from an electrical grid; sometimes, however, they are self-excited by using phase-correcting capacitors. Because of this, induction generators cannot usually "black start" a de-energized distribution system.

Principle of Operation

An induction generator produces electrical power when its rotor is turned faster than the *synchronous speed*. For a typical four-pole motor (two pairs of poles on stator) operating on a 60 Hz electrical grid, the synchronous speed is 1800 rotations per minute (rpm). The same four-pole motor operating on a 50 Hz grid will have a synchronous speed of 1500 RPM. The motor normally turns slightly slower than the synchronous speed; the difference between synchronous and operating speed is called "slip" and is usually expressed as per cent of the synchronous speed. For example, a motor operating at 1450 RPM that has a synchronous speed of 1500 RPM is running at a slip of +3.3%.

In normal motor operation, the stator flux rotation is faster than the rotor rotation. This causes the stator flux to induce rotor currents, which create a rotor flux with magnetic polarity opposite to stator. In this way, the rotor is dragged along behind stator flux, with the currents in the rotor induced at the slip frequency.

In generator operation, a prime mover (turbine or engine) drives the rotor above the synchronous speed (negative slip). The stator flux still induces currents in the rotor, but since the opposing rotor flux is now cutting the stator coils, an active current is produced in stator coils and the motor now operates as a generator, sending power back to the electrical grid.

Excitation

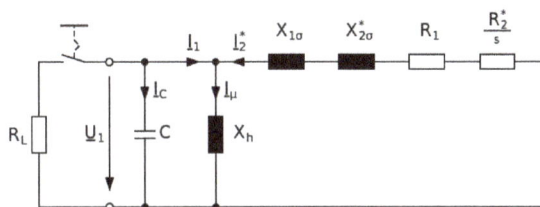

Equivalent circuit of induction generator

An induction machine requires externally supplied armature current; it cannot start on its own as a generator. Because the rotor field always lags behind the stator field, the induction machine always "consumes" reactive power, regardless of whether it is operating as a generator or a motor.

A source of excitation current for magnetizing flux (reactive power) for the stator is still required, to induce rotor current. This can be supplied from the electrical grid or, once it starts producing power, from the generator itself.

Active Power

Active power delivered to the line is proportional to slip above the synchronous speed. Full rated power of the generator is reached at very small slip values (motor dependent, typically 3%). At synchronous speed of 1800 rpm, generator will produce no power. When the driving speed is increased to 1860 rpm (typical example), full output power is produced. If the prime mover is unable to produce enough power to fully drive the generator, speed will remain somewhere between 1800 and 1860 rpm range.

Required Capacitance

A capacitor bank must supply reactive power to the motor when used in stand-alone mode. The reactive power supplied should be equal or greater than the reactive power that the machine normally draws when operating as a motor.

Torque Vs. Slip

The basic fundamental of induction generators is the conversion between mechanical energy to electrical energy. This requires an external torque applied to the rotor to turn it faster than the synchronous speed. However, indefinitely increasing torque doesn't lead to an indefinite increase in power generation. The rotating magnetic field torque excited from the armature works to counter the motion of the rotor and prevent over speed because of induced motion in the opposite direction. As the speed of the motor increases the counter torque reaches a max value of torque (breakdown torque) that it can operate until before the operating conditions become unstable. Ideally, induction generators work best in the stable region between the no-load condition and maximum torque region.

Maximum Pass-Through Current

In practice and without taking this notion into account, many users unsuccessfully apply the principles to the actual deployment.

It's not in popular belief; that in almost every case, under the same active grid voltage, the power that the generator produces is greater than the power it consumes when it is at the motor,fully loaded state; its rated power. Sometimes the differences are in multiple folds. Higher the power means higher the amperage.

For prolong operation, and implied in its guaranteed, each motor has its "maximum pass-through current". This amperage value; the current density; is derived from the maximum pass-through current property of the internal copper magnet wire and the combined configuration of their connections. Without opening up the unit to examine the internal setting of the copper wires, a division of the wattage of its rated power by its rated voltage can give users some senses of how much that value is.

Therefore, claims of making a unit generates more power than its rated should get a closer examination.

Grid and Stand-Alone Connections

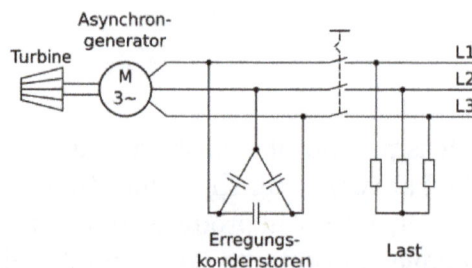

Typical connections when used as a standalone generator

In induction generators, the reactive power required to establish the air gap magnetic flux is provided by a capacitor bank connected to the machine in case of stand-alone system and in case of grid connection it draws reactive power from the grid to maintain its air gap flux. For a grid-connected system, frequency and voltage at the machine will be dictated by the electric grid, since it is very small compared to the whole system. For stand-alone systems, frequency and voltage are complex function of machine parameters, capacitance used for excitation, and load value and type.

Limitations of Induction Generators

An induction generator cannot generate reactive power. Actually it requires reactive power from supply line to furnish its excitation, since it has no means for establishing air gap flux with stator open circuited. Operation of an induction generator requires a synchronous machine, whether generator or motor, on the line to supply the induction generator with its needed excitation power. This requirement limits the use of induction generators to specialty applications.

Use of Induction Generators

Induction generators are often used in wind turbines and some micro hydro installations due to their ability to produce useful power at varying rotor speeds. Induction generators are mechanically and electrically simpler than other generator types. They are also more rugged, requiring no brushes or commutators.

Induction generators are particularly suitable for wind generating stations as in this case speed is always a variable factor. Unlike synchronous motors, induction generators are load-dependent and cannot be used alone for grid frequency control.

Example Application

As an example, consider the use of a 10 hp, 1760 r/min, 440 V, three-phase induction motor as an asynchronous generator. The full-load current of the motor is 10 A and the full-load power factor is 0.8.

Required capacitance per phase if capacitors are connected in delta:

Apparent power S = $\sqrt{3}$ E I = 1.73 × 440 × 10 = 7612 VA

Active power P = S cos θ = 7612 × 0.8 = 6090 W

Reactive power Q = $\sqrt{S^2 - P^2}$ = 4567 VAR

For a machine to run as an asynchronous generator, capacitor bank must supply minimum 4567 / 3 phases = 1523 VAR per phase. Voltage per capacitor is 440 V because capacitors are connected in delta.

Capacitive current Ic = Q/E = 1523/440 = 3.46 A

Capacitive reactance per phase Xc = E/Ic = 127 Ω

Minimum capacitance per phase:

C = 1 / (2*π*f*Xc) = 1 / (2 * 3.141 * 60 * 127) = 21 microfarads.

If the load also absorbs reactive power, capacitor bank must be increased in size to compensate.

Prime mover speed should be used to generate frequency of 60 Hz:

Typically, slip should be similar to full-load value when machine is running as motor, but negative (generator operation):

> if Ns = 1800, one can choose N=Ns+40 rpm

> Required prime mover speed N = 1800 + 40 = 1840 rpm.

Solar Cell

A conventional crystalline silicon solar cell (as of 2005). Electrical contacts made from busbars (the larger silver-colored strips) and fingers (the smaller ones) are printed on the silicon wafer.

A solar cell, or photovoltaic cell (in very early days also termed "solar battery" – a denotation which nowadays has a totally different meaning), is an electrical device that converts the energy of light directly into electricity by the photovoltaic effect, which is a physical and chemical phenomenon. It is a form of photoelectric cell, defined as a device whose electrical characteristics, such as current, voltage, or resistance, vary when exposed to light. Solar cells are the building blocks of photovoltaic modules, otherwise known as solar panels.

Solar cells are described as being photovoltaic irrespective of whether the source is sunlight or an artificial light. They are used as a photodetector (for example infrared detectors), detecting light or other electromagnetic radiation near the visible range, or measuring light intensity.

The operation of a photovoltaic (PV) cell requires 3 basic attributes:

- The absorption of light, generating either electron-hole pairs or excitons.
- The separation of charge carriers of opposite types.
- The separate extraction of those carriers to an external circuit.

In contrast, a solar thermal collector supplies heat by absorbing sunlight, for the purpose of either direct heating or indirect electrical power generation from heat. A "photoelectrolytic cell" (photo-electrochemical cell), on the other hand, refers either to a type of photovoltaic cell (like that developed by Edmond Becquerel and modern dye-sensitized solar cells), or to a device that splits water directly into hydrogen and oxygen using only solar illumination.

Applications

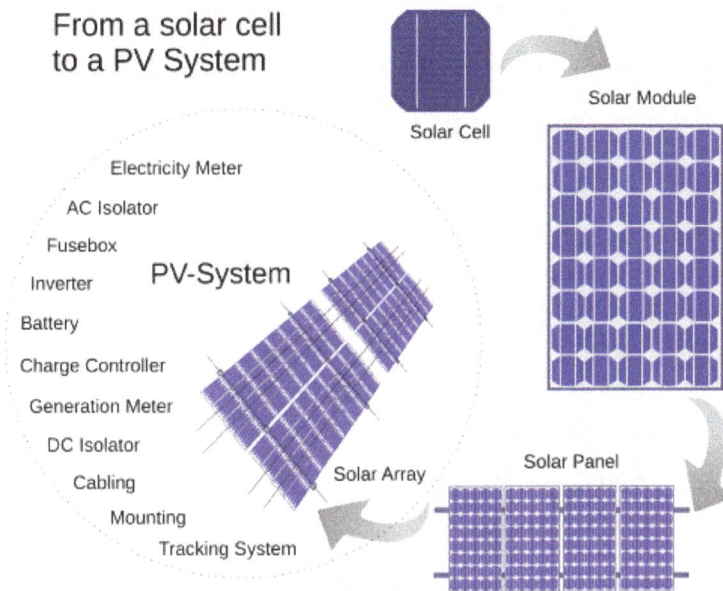

From a solar cell to a PV system. Diagram of the possible components of a photovoltaic system

Assemblies of solar cells are used to make solar modules which generate electrical power from sunlight, as distinguished from a "solar thermal module" or "solar hot water panel". A solar array generates solar power using solar energy.

Cells, Modules, Panels and Systems

Multiple solar cells in an integrated group, all oriented in one plane, constitute a solar photovoltaic panel or solar photovoltaic module. Photovoltaic modules often have a sheet of glass on the sun-facing side, allowing light to pass while protecting the semiconductor wafers. Solar cells are usually connected in series and parallel circuits or series in modules, creating an additive voltage. Connecting cells in parallel yields a higher current; however, problems such as shadow effects can shut down the weaker (less illuminated) parallel string (a number of series connected cells) causing substantial power loss and possible damage because of the reverse bias applied to the shadowed cells by their illuminated partners. Strings of series cells are usually handled independently and not connected in parallel, though as of 2014, individual power boxes are often supplied for each module, and are connected in parallel. Although modules can be interconnected to create an array with the desired peak DC voltage and loading current capacity, using independent MPPTs (maximum power point trackers) is preferable. Otherwise, shunt diodes can reduce shadowing power loss in arrays with series/parallel connected cells.

Typical PV system prices in 2013 in selected countries (USD)								
USD/W	Australia	China	France	Germany	Italy	Japan	United Kingdom	United States
Residential	1.8	1.5	4.1	2.4	2.8	4.2	2.8	4.9
Commercial	1.7	1.4	2.7	1.8	1.9	3.6	2.4	4.5
Utility-scale	2.0	1.4	2.2	1.4	1.5	2.9	1.9	3.3

Source: IEA – Technology Roadmap: Solar Photovoltaic Energy report, 2014 edition
Note: DOE – Photovoltaic System Pricing Trends reports lower prices for the U.S.

History

The photovoltaic effect was experimentally demonstrated first by French physicist Edmond Becquerel. In 1839, at age 19, he built the world's first photovoltaic cell in his father's laboratory. Willoughby Smith first described the "Effect of Light on Selenium during the passage of an Electric Current" in a 20 February 1873 issue of Nature. In 1883 Charles Fritts built the first solid state photovoltaic cell by coating the semiconductor selenium with a thin layer of gold to form the junctions; the device was only around 1% efficient.

In 1888 Russian physicist Aleksandr Stoletov built the first cell based on the outer photoelectric effect discovered by Heinrich Hertz in 1887.

In 1905 Albert Einstein proposed a new quantum theory of light and explained the photoelectric effect in a landmark paper, for which he received the Nobel Prize in Physics in 1921.

Vadim Lashkaryov discovered p-n-junctions in CuO and silver sulphide protocells in 1941.

Russell Ohl patented the modern junction semiconductor solar cell in 1946 while working on the series of advances that would lead to the transistor.

The first practical photovoltaic cell was publicly demonstrated on 25 April 1954 at Bell Laboratories. The inventors were Daryl Chapin, Calvin Souther Fuller and Gerald Pearson.

Solar cells gained prominence with their incorporation onto the 1958 Vanguard I satellite.

Improvements were gradual over the next two decades. However, this success was also the reason that costs remained high, because space users were willing to pay for the best possible cells, leaving no reason to invest in lower-cost, less-efficient solutions. The price was determined largely by the semiconductor industry; their move to integrated circuits in the 1960s led to the availability of larger boules at lower relative prices. As their price fell, the price of the resulting cells did as well. These effects lowered 1971 cell costs to some $100 per watt.

Space Applications

Solar cells were first used in a prominent application when they were proposed and flown on the Vanguard satellite in 1958, as an alternative power source to the primary battery power source. By adding cells to the outside of the body, the mission time could be extended with no major changes to the spacecraft or its power systems. In 1959 the United States launched Explorer 6, featuring large wing-shaped solar arrays, which became a common feature in satellites. These arrays consisted of 9600 Hoffman solar cells.

By the 1960s, solar cells were (and still are) the main power source for most Earth orbiting satellites and a number of probes into the solar system, since they offered the best power-to-weight ratio. However, this success was possible because in the space application, power system costs could be high, because space users had few other power options, and were willing to pay for the best possible cells. The space power market drove the development of higher efficiencies in solar cells up until the National Science Foundation "Research Applied to National Needs" program began to push development of solar cells for terrestrial applications.

In the early 1990s the technology used for space solar cells diverged from the silicon technology used for terrestrial panels, with the spacecraft application shifting to gallium arsenide-based III-V semiconductor materials, which then evolved into the modern III-V multijunction photovoltaic cell used on spacecraft.

Price Reductions

In late 1969 Elliot Berman joined Exxon's task force which was looking for projects 30 years in the future and in April 1973 he founded Solar Power Corporation, a wholly owned subsidiary of Exxon at that time. The group had concluded that electrical power would be much more expensive by 2000, and felt that this increase in price would make alternative energy sources more attractive. He conducted a market study and concluded that a price per watt of about $20/watt would create significant demand. The team eliminated the steps of polishing the wafers and coating them with an anti-reflective layer, relying on the rough-sawn wafer surface. The team also replaced the expensive materials and hand wiring used in space applications with a printed circuit board on the back, acrylic plastic on the front, and silicone glue between the two, "potting" the cells. Solar cells could be made using cast-off material from the electronics market. By 1973 they announced a product, and SPC convinced Tideland Signal to use its panels to power navigational buoys, initially for the U.S. Coast Guard.

Research into solar power for terrestrial applications became prominent with the U.S. National Science Foundation's Advanced Solar Energy Research and Development Division within the "Research Applied to National Needs" program, which ran from 1969 to 1977, and funded research on developing solar power for ground electrical power systems. A 1973 conference, the "Cherry Hill Conference", set forth the technology goals required to achieve this goal and outlined an ambitious project for achieving them, kicking off an applied research program that would be ongoing for several decades. The program was eventually taken over by the Energy Research and Development Administration (ERDA), which was later merged into the U.S. Department of Energy.

Following the 1973 oil crisis, oil companies used their higher profits to start (or buy) solar firms, and were for decades the largest producers. Exxon, ARCO, Shell, Amoco (later purchased by BP) and Mobil all had major solar divisions during the 1970s and 1980s. Technology companies also participated, including General Electric, Motorola, IBM, Tyco and RCA.

Declining Costs and Exponential Growth

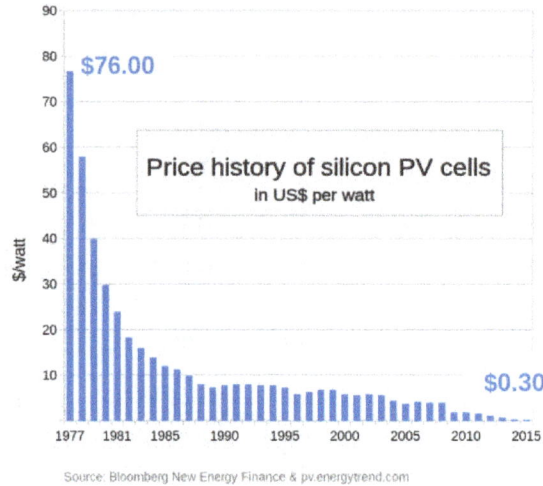

Price per watt history for conventional (c-Si) solar cells since 1977

Adjusting for inflation, it cost $96 per watt for a solar module in the mid-1970s. Process improvements and a very large boost in production have brought that figure down 99 percent, to 68¢ per watt in 2016, according to data from Bloomberg New Energy Finance. Swanson's law is an observation similar to Moore's Law that states that solar cell prices fall 20% for every doubling of industry capacity. It was featured in an article in the British weekly newspaper The Economist.

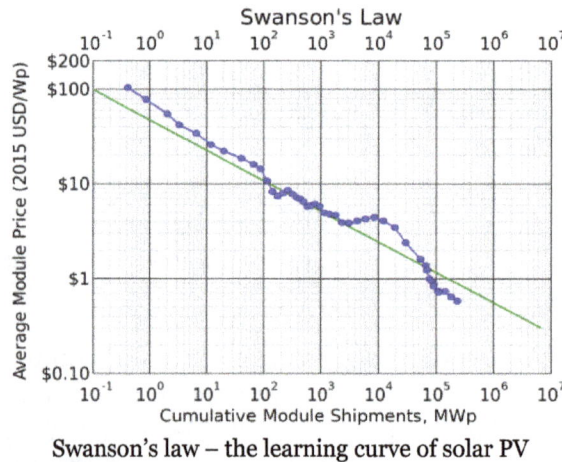

Swanson's law – the learning curve of solar PV

Further improvements reduced production cost to under $1 per watt, with wholesale costs well under $2. Balance of system costs were then higher than the panels. Large commercial arrays could be built, as of 2010, at below $3.40 a watt, fully commissioned.

As the semiconductor industry moved to ever-larger boules, older equipment became inexpensive. Cell sizes grew as equipment became available on the surplus market; ARCO Solar's original panels used cells 2 to 4 inches (50 to 100 mm) in diameter. Panels in the 1990s and early 2000s generally used 125 mm wafers; since 2008 almost all new panels use 150 mm cells. The widespread introduction of flat screen televisions in the late 1990s and early 2000s led to the wide availability of large, high-quality glass sheets to cover the panels.

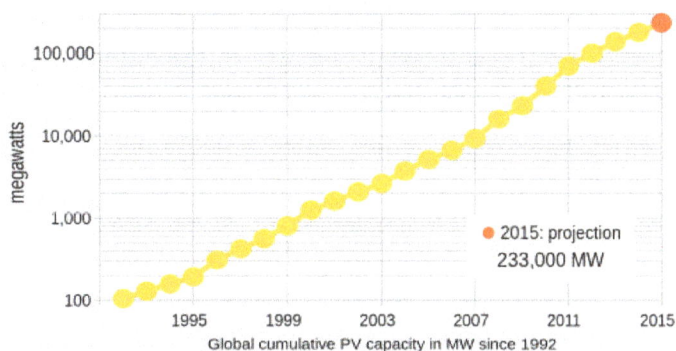

Growth of photovoltaics – Worldwide total installed PV capacity

During the 1990s, polysilicon ("poly") cells became increasingly popular. These cells offer less efficiency than their monosilicon ("mono") counterparts, but they are grown in large vats that reduce cost. By the mid-2000s, poly was dominant in the low-cost panel market, but more recently the mono returned to widespread use.

Manufacturers of wafer-based cells responded to high silicon prices in 2004–2008 with rapid reductions in silicon consumption. In 2008, according to Jef Poortmans, director of IMEC's organic and solar department, current cells use 8–9 grams (0.28–0.32 oz) of silicon per watt of power generation, with wafer thicknesses in the neighborhood of 200 microns.

First Solar is the largest thin film manufacturer in the world, using a CdTe-cell sandwiched between two layers of glass. Crystalline silicon panels dominate worldwide markets and are mostly manufactured in China and Taiwan. By late 2011, a drop in European demand due to budgetary turmoil dropped prices for crystalline solar modules to about $1.09 per watt down sharply from 2010. Prices continued to fall in 2012, reaching $0.62/watt by 4Q2012.

Global installed PV capacity reached at least 177 gigawatts in 2014, enough to supply 1 percent of the world's total electricity consumption. Solar PV is growing fastest in Asia, with China and Japan currently accounting for half of worldwide deployment.

Subsidies and Grid Parity

Solar-specific feed-in tariffs vary by country and within countries. Such tariffs encourage the development of solar power projects. Widespread grid parity, the point at which photovoltaic electricity is equal to or cheaper than grid power without subsidies, likely requires advances on all three fronts. Proponents of solar hope to achieve grid parity first in areas with abundant sun and high electricity costs such as in California and Japan. In 2007 BP claimed grid parity for Hawaii and other islands that otherwise use diesel fuel to produce electricity. George W. Bush set 2015 as the date for grid parity in the US. The Photovoltaic Association reported in 2012 that Australia had reached grid parity (ignoring feed in tariffs).

The price of solar panels fell steadily for 40 years, interrupted in 2004 when high subsidies in Germany drastically increased demand there and greatly increased the price of purified silicon (which is used in computer chips as well as solar panels). The recession of 2008 and the onset of Chinese manufacturing caused prices to resume their decline. In the four years after January 2008 prices for solar modules in Germany dropped from €3 to €1 per peak watt. During that same time production capacity surged with an annual growth of more than 50%. China increased market share from 8% in 2008 to over 55% in the last quarter of 2010. In December 2012 the price of Chinese solar panels had dropped to $0.60/Wp (crystalline modules).

Theory

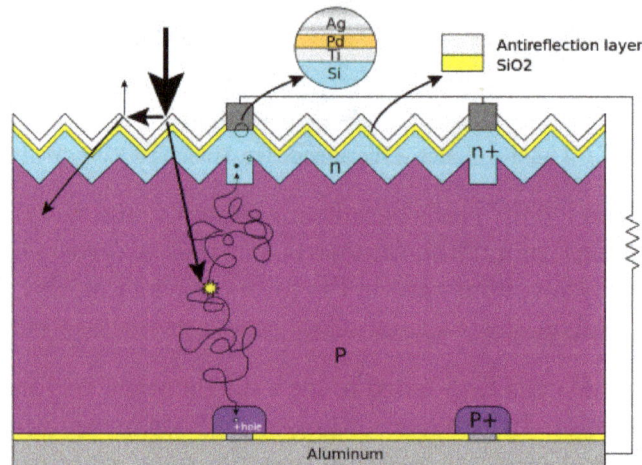

Working mechanism of a solar cell

The solar cell works in several steps:

- Photons in sunlight hit the solar panel and are absorbed by semiconducting materials, such as silicon.

- Electrons are excited from their current molecular/atomic orbital. Once excited an electron can either dissipate the energy as heat and return to its orbital or travel through the cell until it reaches an electrode. Current flows through the material to cancel the potential and this electricity is captured. The chemical bonds of the material are vital for this process to work, and usually silicon is used in two layers, one layer being bonded with boron, the other phosphorus. These layers have different chemical electric charges and subsequently both drive and direct the current of electrons.

- An array of solar cells converts solar energy into a usable amount of direct current (DC) electricity.

- An inverter can convert the power to alternating current (AC).

The most commonly known solar cell is configured as a large-area p–n junction made from silicon.

Efficiency

Solar cell efficiency may be broken down into reflectance efficiency, thermodynamic efficiency, charge carrier separation efficiency and conductive efficiency. The overall efficiency is the product of these individual metrics.

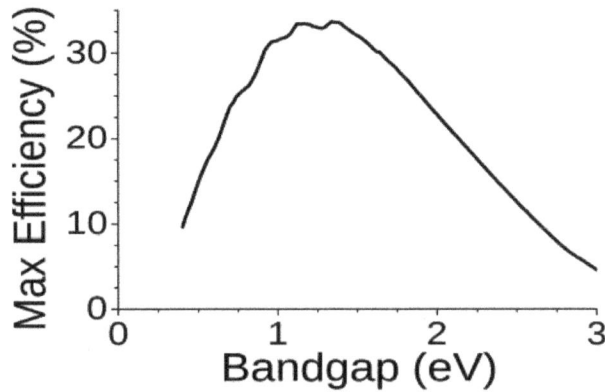

The Shockley-Queisser limit for the theoretical maximum efficiency of a solar cell. Semiconductors with band gap between 1 and 1.5eV, or near-infrared light, have the greatest potential to form an efficient single-junction cell. (The efficiency "limit" shown here can be exceeded by multijunction solar cells.)

A solar cell has a voltage dependent efficiency curve, temperature coefficients, and allowable shadow angles.

Due to the difficulty in measuring these parameters directly, other parameters are substituted: thermodynamic efficiency, quantum efficiency, integrated quantum efficiency, V_{OC} ratio, and fill factor. Reflectance losses are a portion of quantum efficiency under "external quantum efficiency". Recombination losses make up another portion of quantum efficiency, V_{OC} ratio, and fill factor. Resistive losses are predominantly categorized under fill factor, but also make up minor portions of quantum efficiency, V_{OC} ratio.

The fill factor is the ratio of the actual maximum obtainable power to the product of the open circuit voltage and short circuit current. This is a key parameter in evaluating performance. In 2009, typical commercial solar cells had a fill factor > 0.70. Grade B cells were usually between 0.4 and 0.7. Cells with a high fill factor have a low equivalent series resistance and a high equivalent shunt resistance, so less of the current produced by the cell is dissipated in internal losses.

Single p–n junction crystalline silicon devices are now approaching the theoretical limiting power efficiency of 33.7%, noted as the Shockley–Queisser limit in 1961. In the extreme, with an infinite number of layers, the corresponding limit is 86% using concentrated sunlight.

In December 2014, a solar cell achieved a new laboratory record with 46 percent efficiency in a French-German collaboration.

In 2014, three companies broke the record of 25.6% for a silicon solar cell. Panasonic's was the most efficient. The company moved the front contacts to the rear of the panel, eliminating shaded areas. In addition they applied thin silicon films to the (high quality silicon) wafer's front and back to eliminate defects at or near the wafer surface.

In September 2015, the Fraunhofer Institute for Solar Energy Systems (Fraunhofer ISE) announced the achievement of an efficiency above 20% for epitaxial wafer cells. The work on optimizing the atmospheric-pressure chemical vapor deposition (APCVD) in-line production chain was done in collaboration with NexWafe GmbH, a company spun off from Fraunhofer ISE to commercialize production.

For triple-junction thin-film solar cells, the world record is 13.6%, set in June 2015.

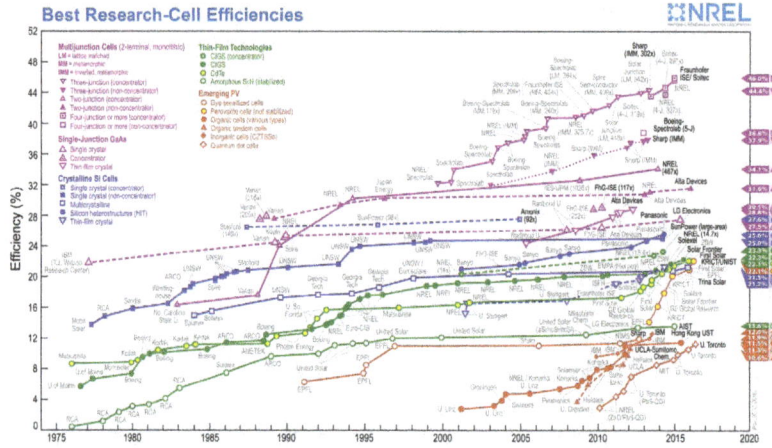

Reported timeline of solar cell energy conversion efficiencies (National Renewable Energy Laboratory)

Materials

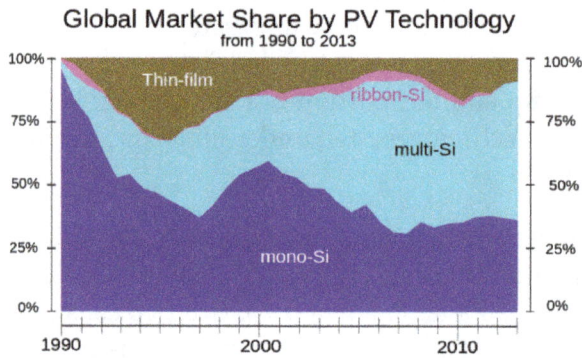

Global market-share in terms of annual production by PV technology since 1990

Solar cells are typically named after the semiconducting material they are made of. These materials must have certain characteristics in order to absorb sunlight. Some cells are designed to handle sunlight that reaches the Earth's surface, while others are optimized for use in space. Solar cells can be made of only one single layer of light-absorbing material (single-junction) or use multiple physical configurations (multi-junctions) to take advantage of various absorption and charge separation mechanisms.

Solar cells can be classified into first, second and third generation cells. The first generation cells—also called conventional, traditional or wafer-based cells—are made of crystalline silicon, the commercially predominant PV technology, that includes materials such as polysilicon and monocrystalline silicon. Second generation cells are thin film solar cells, that include amorphous silicon, CdTe and CIGS cells and are commercially significant in utility-scale photovoltaic power stations, building integrated photovoltaics or in small stand-alone power system. The third generation of solar cells includes a number of thin-film technologies often described as emerging photovoltaics—most of them have not yet been commercially applied and are still in the research or development phase. Many use organic materials, often organometallic compounds as well as inorganic substances. Despite the fact that their efficiencies had been low and the stability of the absorber material was often too short for commercial applications, there is a lot of research invested into these technologies as they promise to achieve the goal of producing low-cost, high-efficiency solar cells.

Crystalline Silicon

By far, the most prevalent bulk material for solar cells is crystalline silicon (c-Si), also known as "solar grade silicon". Bulk silicon is separated into multiple categories according to crystallinity and crystal size in the resulting ingot, ribbon or wafer. These cells are entirely based around the concept of a p-n junction. Solar cells made of c-Si are made from wafers between 160 and 240 micrometers thick.

Monocrystalline Silicon

Monocrystalline silicon (mono-Si) solar cells are more efficient and more expensive than most other types of cells. The corners of the cells look clipped, like an octagon, because the wafer material is cut from cylindrical ingots, that are typically grown by the Czochralski process. Solar panels using mono-Si cells display a distinctive pattern of small white diamonds.

Epitaxial Silicon

Epitaxial wafers can be grown on a monocrystalline silicon "seed" wafer by atmospheric-pressure CVD in a high-throughput inline process, and then detached as self-supporting wafers of some standard thickness (e.g., 250 μm) that can be manipulated by hand, and directly substituted for wafer cells cut from monocrystalline silicon ingots. Solar cells made with this technique can have efficiencies approaching those of wafer-cut cells, but at appreciably lower cost.

Polycrystalline Silicon

Polycrystalline silicon, or multicrystalline silicon (multi-Si) cells are made from cast square ingots—large blocks of molten silicon carefully cooled and solidified. They consist of small crystals giving the material its typical metal flake effect. Polysilicon cells are the most common type used in photovoltaics and are less expensive, but also less efficient, than those made from monocrystalline silicon.

Ribbon Silicon

Ribbon silicon is a type of polycrystalline silicon—it is formed by drawing flat thin films from molten silicon and results in a polycrystalline structure. These cells are cheaper to make than multi-Si, due to a great reduction in silicon waste, as this approach does not require sawing from ingots. However, they are also less efficient.

Mono-Like-Multi Silicon (MLM)

This form was developed in the 2000s and introduced commercially around 2009. Also called cast-mono, this design uses polycrystalline casting chambers with small "seeds" of mono material. The result is a bulk mono-like material that is polycrystalline around the outsides. When sliced for processing, the inner sections are high-efficiency mono-like cells (but square instead of "clipped"), while the outer edges are sold as conventional poly. This production method results in mono-like cells at poly-like prices.

Thin Film

Thin-film technologies reduce the amount of active material in a cell. Most designs sandwich active material between two panes of glass. Since silicon solar panels only use one pane of glass, thin film panels are approximately twice as heavy as crystalline silicon panels, although they have a smaller ecological impact (determined from life cycle analysis). The majority of film panels have 2–3 percentage points lower conversion efficiencies than crystalline silicon. Cadmium telluride (CdTe), copper indium gallium selenide (CIGS) and amorphous silicon (a-Si) are three thin-film technologies often used for outdoor applications. As of December 2013, CdTe cost per installed watt was \$0.59 as reported by First Solar. CIGS technology laboratory demonstrations reached 20.4% conversion efficiency as of December 2013. The lab efficiency of GaAs thin film technology topped 28%. The quantum efficiency of thin film solar cells is also lower due to reduced number of collected charge carriers per incident photon. Most recently, CZTS solar cell emerge as the less-toxic thin film solar cell technology, which achieved ~12% efficiency.

Cadmium Telluride

Cadmium telluride is the only thin film material so far to rival crystalline silicon in cost/watt. However cadmium is highly toxic and tellurium (anion: "telluride") supplies are limited. The cadmium present in the cells would be toxic if released. However, release is impossible during normal operation of the cells and is unlikely during fires in residential roofs. A square meter of CdTe contains approximately the same amount of Cd as a single C cell nickel-cadmium battery, in a more stable and less soluble form.

Copper Indium Gallium Selenide

Copper indium gallium selenide (CIGS) is a direct band gap material. It has the highest efficiency (~20%) among all commercially significant thin film materials. Traditional methods of fabrication involve vacuum processes including co-evaporation and sputtering. Recent developments at IBM and Nanosolar attempt to lower the cost by using non-vacuum solution processes.

Silicon Thin Film

Silicon thin-film cells are mainly deposited by chemical vapor deposition (typically plasma-enhanced, PE-CVD) from silane gas and hydrogen gas. Depending on the deposition parameters, this can yield amorphous silicon (a-Si or a-Si:H), protocrystalline silicon or nanocrystalline silicon (nc-Si or nc-Si:H), also called microcrystalline silicon.

Amorphous silicon is the most well-developed thin film technology to-date. An amorphous silicon (a-Si) solar cell is made of non-crystalline or microcrystalline silicon. Amorphous silicon has a higher bandgap (1.7 eV) than crystalline silicon (c-Si) (1.1 eV), which means it absorbs the visible part of the solar spectrum more strongly than the higher power density infrared portion of the spectrum. The production of a-Si thin film solar cells uses glass as a substrate and deposits a very thin layer of silicon by plasma-enhanced chemical vapor deposition (PECVD).

Protocrystalline silicon with a low volume fraction of nanocrystalline silicon is optimal for high open circuit voltage. Nc-Si has about the same bandgap as c-Si and nc-Si and a-Si can advantageously be combined in thin layers, creating a layered cell called a tandem cell. The top cell in a-Si absorbs the visible light and leaves the infrared part of the spectrum for the bottom cell in nc-Si.

Gallium Arsenide Thin Film

The semiconductor material Gallium arsenide (GaAs) is also used for single-crystalline thin film solar cells. Although GaAs cells are very expensive, they hold the world's record in efficiency for a single-junction solar cell at 28.8%. GaAs is more commonly used in multijunction photovoltaic cells for concentrated photovoltaics (CPV, HCPV) and for solar panels on spacecrafts, as the industry favours efficiency over cost for space-based solar power.

Multijunction Cells

Dawn's 10 kW triple-junction gallium arsenide solar array at full extension

Multi-junction cells consist of multiple thin films, each essentially a solar cell grown on top of another, typically using metalorganic vapour phase epitaxy. Each layers has a different band gap energy to allow it to absorb electromagnetic radiation over a different portion of the spectrum. Multi-junction cells were originally developed for special applications such as satellites and space exploration, but are now used increasingly in terrestrial concentrator photovoltaics (CPV), an emerging technology that uses lenses and curved mirrors to concentrate sunlight onto small but highly efficient multi-junction solar cells. By concentrating sunlight up to a thousand times, *High concentrated photovoltaics (HCPV)* has the potential to outcompete conventional solar PV in the future.

Tandem solar cells based on monolithic, series connected, gallium indium phosphide (GaInP), gallium arsenide (GaAs), and germanium (Ge) p–n junctions, are increasing sales, despite cost pressures. Between December 2006 and December 2007, the cost of 4N gallium metal rose from about $350 per kg to $680 per kg. Additionally, germanium metal prices have risen substantially to $1000–1200 per kg this year. Those materials include gallium (4N, 6N and 7N Ga), arsenic (4N, 6N and 7N) and germanium, pyrolitic boron nitride (pBN) crucibles for growing crystals, and boron oxide, these products are critical to the entire substrate manufacturing industry.

A triple-junction cell, for example, may consist of the semiconductors: GaAs, Ge, and GaInP 2. Triple-junction GaAs solar cells were used as the power source of the Dutch four-time World Solar Challenge winners Nuna in 2003, 2005 and 2007 and by the Dutch solar cars Solutra (2005), Twente One (2007) and 21Revolution (2009). GaAs based multi-junction devices are the most efficient solar cells to date. On 15 October 2012, triple junction metamorphic cells reached a record high of 44%.

Research in Solar Cells

Perovskite Solar Cells

Perovskite solar cells are solar cells that include a perovskite-structured material as the active layer. Most commonly, this is a solution-processed hybrid organic-inorganic tin or lead halide based material. Efficiencies have increased from below 5% at their first usage in 2009 to over 20% in 2014, making them a very rapidly advancing technology and a hot topic in the solar cell field. Perovskite solar cells are also forecast to be extremely cheap to scale up, making them a very attractive option for commercialisation.

Liquid Inks

In 2014, researchers at California NanoSystems Institute discovered using kesterite and perovskite improved electric power conversion efficiency for solar cells.

Upconversion and Downconversion

Photon upconversion is the process of using two low-energy (*e.g.*, infrared) photons to produce one higher energy photon; downconversion is the process of using one high energy photon (*e.g.*, ultraviolet) to produce two lower energy photons. Either of these techniques could be used to produce higher efficiency solar cells by allowing solar photons to be more efficiently used. The difficulty, however, is that the conversion efficiency of existing phosphors exhibiting up- or down-conversion is low, and is typically narrow band.

One upconversion technique is to incorporate lanthanide-doped materials ($Er3+$, $Yb3+$, $Ho3+$or a combination), taking advantage of their luminescence to convert infrared radiation to visible light. Upconversion process occurs when two infrared photons are absorbed by rare-earth ions to generate a (high-energy) absorbable photon. As example, the energy transfer upconversion process (ETU), consists in successive transfer processes between excited ions in the near infrared. The upconverter material could be placed below the solar cell to absorb the infrared light that passes through the silicon. Useful ions are most commonly found in the trivalent state. $Er+$ions have been the most used. $Er3+$ions absorb solar radiation around 1.54 µm. Two $Er3+$ions that have absorbed this radiation can interact with each other through an upconversion process. The excited ion emits light above the Si bandgap that is absorbed by the solar cell and creates an additional electron–hole pair that can generate current. However, the increased efficiency was small. In addition, fluoroindate glasses have low phonon energy and have been proposed as suitable matrix doped with $Ho3+$ions.

Light-Absorbing Dyes

Dye-sensitized solar cells (DSSCs) are made of low-cost materials and do not need elaborate manufacturing equipment, so they can be made in a DIY fashion. In bulk it should be significantly less expensive than older solid-state cell designs. DSSC's can be engineered into flexible sheets and although its conversion efficiency is less than the best thin film cells, its price/performance ratio may be high enough to allow them to compete with fossil fuel electrical generation.

Typically a ruthenium metalorganic dye (Ru-centered) is used as a monolayer of light-absorbing material. The dye-sensitized solar cell depends on a mesoporous layer of nanoparticulate titanium dioxide to greatly amplify the surface area (200–300 m²/g TiO2, as compared to approximately 10 m²/g of flat single crystal). The photogenerated electrons from the light absorbing dye are passed on to the n-type TiO2 and the holes are absorbed by an electrolyte on the other side of the dye. The circuit is completed by a redox couple in the electrolyte, which can be liquid or solid. This type of cell allows more flexible use of materials and is typically manufactured by screen printing or ultrasonic nozzles, with the potential for lower processing costs than those used for bulk solar cells. However, the dyes in these cells also suffer from degradation under heat and UV light and the cell casing is difficult to seal due to the solvents used in assembly. The first commercial shipment of DSSC solar modules occurred in July 2009 from G24i Innovations.

Quantum Dots

Quantum dot solar cells (QDSCs) are based on the Gratzel cell, or dye-sensitized solar cell architecture, but employ low band gap semiconductor nanoparticles, fabricated with crystallite sizes small enough to form quantum dots (such as CdS, CdSe, Sb2S3, PbS, etc.), instead of organic or organometallic dyes as light absorbers. QD's size quantization allows for the band gap to be tuned by simply changing particle size. They also have high extinction coefficients and have shown the possibility of multiple exciton generation.

In a QDSC, a mesoporous layer of titanium dioxide nanoparticles forms the backbone of the cell, much like in a DSSC. This TiO2 layer can then be made photoactive by coating with semiconductor quantum dots using chemical bath deposition, electrophoretic deposition or successive ionic layer adsorption and reaction. The electrical circuit is then completed through the use of a liquid or solid redox couple. The efficiency of QDSCs has increased to over 5% shown for both liquid-junction and solid state cells. In an effort to decrease production costs, the Prashant Kamat research group demonstrated a solar paint made with TiO2 and CdSe that can be applied using a one-step method to any conductive surface with efficiencies over 1%.

Organic/Polymer Solar Cells

Organic solar cells and polymer solar cells are built from thin films (typically 100 nm) of organic semiconductors including polymers, such as polyphenylene vinylene and small-molecule compounds like copper phthalocyanine (a blue or green organic pigment) and carbon fullerenes and fullerene derivatives such as PCBM.

They can be processed from liquid solution, offering the possibility of a simple roll-to-roll printing process, potentially leading to inexpensive, large-scale production. In addition, these cells could be beneficial for some applications where mechanical flexibility and disposability are important. Current cell efficiencies are, however, very low, and practical devices are essentially non-existent.

Energy conversion efficiencies achieved to date using conductive polymers are very low compared to inorganic materials. However, Konarka Power Plastic reached efficiency of 8.3% and organic tandem cells in 2012 reached 11.1%.

The active region of an organic device consists of two materials, one electron donor and one electron acceptor. When a photon is converted into an electron hole pair, typically in the donor material, the charges tend to remain bound in the form of an exciton, separating when the exciton diffuses to the donor-acceptor interface, unlike most other solar cell types. The short exciton diffusion lengths of most polymer systems tend to limit the efficiency of such devices. Nanostructured interfaces, sometimes in the form of bulk heterojunctions, can improve performance.

In 2011, MIT and Michigan State researchers developed solar cells with a power efficiency close to 2% with a transparency to the human eye greater than 65%, achieved by selectively absorbing the ultraviolet and near-infrared parts of the spectrum with small-molecule compounds. Researchers at UCLA more recently developed an analogous polymer solar cell, following the same approach, that is 70% transparent and has a 4% power conversion efficiency. These lightweight, flexible cells can be produced in bulk at a low cost and could be used to create power generating windows.

In 2013, researchers announced polymer cells with some 3% efficiency. They used block copolymers, self-assembling organic materials that arrange themselves into distinct layers. The research focused on P3HT-b-PFTBT that separates into bands some 16 nanometers wide.

Adaptive Cells

Adaptive cells change their absorption/reflection characteristics depending to respond to environmental conditions. An adaptive material responds to the intensity and angle of incident light. At the part of the cell where the light is most intense, the cell surface changes from reflective to adaptive, allowing the light to penetrate the cell. The other parts of the cell remain reflective increasing the retention of the absorbed light within the cell.

In 2014 a system that combined an adaptive surface with a glass substrate that redirect the absorbed to a light absorber on the edges of the sheet. The system also included an array of fixed lenses/mirrors to concentrate light onto the adaptive surface. As the day continues, the concentrated light moves along the surface of the cell. That surface switches from reflective to adaptive when the light is most concentrated and back to reflective after the light moves along.

Manufacture

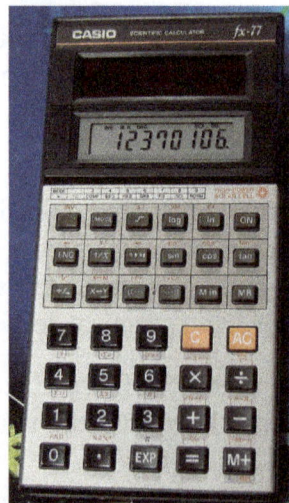

Early solar-powered calculator

Solar cells share some of the same processing and manufacturing techniques as other semiconductor devices. However, the stringent requirements for cleanliness and quality control of semiconductor fabrication are more relaxed for solar cells, lowering costs.

Polycrystalline silicon wafers are made by wire-sawing block-cast silicon ingots into 180 to 350 micrometer wafers. The wafers are usually lightly p-type-doped. A surface diffusion of n-type dopants is performed on the front side of the wafer. This forms a p–n junction a few hundred nanometers below the surface.

Anti-reflection coatings are then typically applied to increase the amount of light coupled into the solar cell. Silicon nitride has gradually replaced titanium dioxide as the preferred material, because of its excellent surface passivation qualities. It prevents carrier recombination at the cell surface. A layer several hundred nanometers thick is applied using PECVD. Some solar cells have textured front surfaces that, like anti-reflection coatings, increase the amount of light reaching the wafer. Such surfaces were first applied to single-crystal silicon, followed by multicrystalline silicon somewhat later.

A full area metal contact is made on the back surface, and a grid-like metal contact made up of fine "fingers" and larger "bus bars" are screen-printed onto the front surface using a silver paste. This is an evolution of the so-called "wet" process for applying electrodes, first described in a US patent filed in 1981 by Bayer AG. The rear contact is formed by screen-printing a metal paste, typically aluminium. Usually this contact covers the entire rear, though some designs employ a grid pattern. The paste is then fired at several hundred degrees Celsius to form metal electrodes in ohmic contact with the silicon. Some companies use an additional electro-plating step to increase efficiency. After the metal contacts are made, the solar cells are interconnected by flat wires or metal ribbons, and assembled into modules or "solar panels". Solar panels have a sheet of tempered glass on the front, and a polymer encapsulation on the back.

Manufacturers and Certification

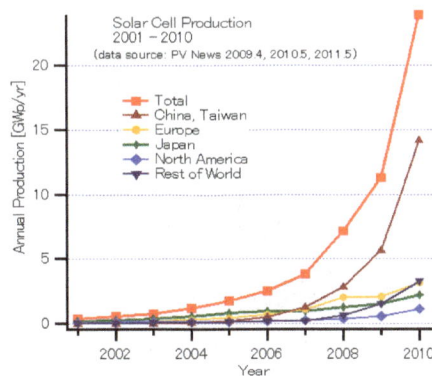

Solar cell production by region

National Renewable Energy Laboratory tests and validates solar technologies. Three reliable groups certify solar equipment: UL and IEEE (both U.S. standards) and IEC.

Solar cells are manufactured in volume in Japan, Germany, China, Taiwan, Malaysia and the United States, whereas Europe, China, the U.S., and Japan have dominated (94% or more as of 2013) in installed systems. Other nations are acquiring significant solar cell production capacity.

Global PV cell/module production increased by 10% in 2012 despite a 9% decline in solar energy investments according to the annual "PV Status Report" released by the European Commission's Joint Research Centre. Between 2009 and 2013 cell production has quadrupled.

China

Due to heavy government investment, China has become the dominant force in solar cell manufacturing. Chinese companies produced solar cells/modules with a capacity of ~23 GW in 2013 (60% of global production).

Malaysia

Photovoltaics manufacturing in Malaysia

In 2014, Malaysia was the world's third largest manufacturer of photovoltaics equipment, behind China and the European Union.

United States

Solar power in the United States

Solar cell production in the U.S. has suffered due to the global financial crisis, but recovered partly due to the falling price of quality silicon.

Tidal Stream Generator

Evopod - A semi-submerged floating approach tested in Strangford Lough.

A tidal stream generator, often referred to as a tidal energy converter (TEC) is a machine that extracts energy from moving masses of water, in particular tides, although the term is often used in reference to machines designed to extract energy from run of river or tidal estuarine sites. Certain types of these machines function very much like underwater wind turbines, and are thus often referred to as tidal turbines. They were first conceived in the 1970s during the oil crisis.

Tidal stream generators are the cheapest and the least ecologically damaging among the three main forms of tidal power generation.

Similarity to Wind Turbines

Tidal stream generators draw energy from water currents in much the same way as wind turbines draw energy from air currents. However, the potential for power generation by an individual tidal turbine can be greater than that of similarly rated wind energy turbine. The higher density of water relative to air (water is about 800 times the density of air) means that a single generator can provide significant power at low tidal flow velocities compared with similar wind speed. Given that power varies with the density of medium and the cube of velocity, water speeds of nearly one-tenth the speed of wind provide the same power for the same size of turbine system; however this limits the application in practice to places where tide speed is at least 2 knots (1 m/s) even close to neap tides. Furthermore, at higher speeds in a flow between 2 and 3 metres per second in seawater a tidal turbine can typically access four times as much energy per rotor swept area as a similarly rated power wind turbine.

Types of Tidal Stream Generators

No standard tidal stream generator has emerged as the clear winner, among a large variety of designs. Several prototypes have shown promise with many companies making bold claims, some of which are yet to be independently verified, but they have not operated commercially for extended periods to establish performances and rates of return on investments.

The European Marine Energy Centre recognizes six principal types of tidal energy converter. They are horizontal axis turbines, vertical axis turbines, oscillating hydrofoils, venturi devices, Archimedes screws and tidal kites.

Axial Turbines

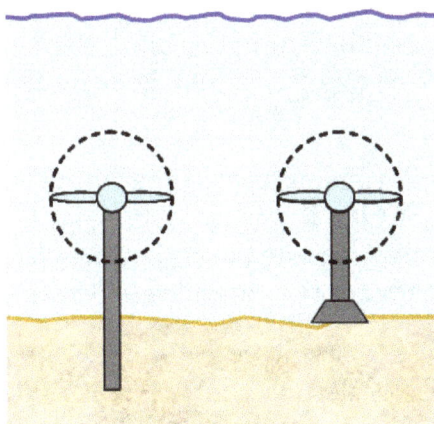

Bottom-mounted axial turbines

These are close in concept to traditional windmills, but operating under the sea. They have most of the prototypes currently operating, including:

Tocardo, a Dutch-based company, has been running tidal turbines since 2008 on the Afsluitdijk, near Den Oever. Typical production data of tidal generator shown of the T100 model as applied in Den Oever. Currently 1 River model (R1) and 2 tidal models (T) are in production with a 3rd T3 coming soon. Power production for the T1 is around 100 kW and around 200 kW for the T2.

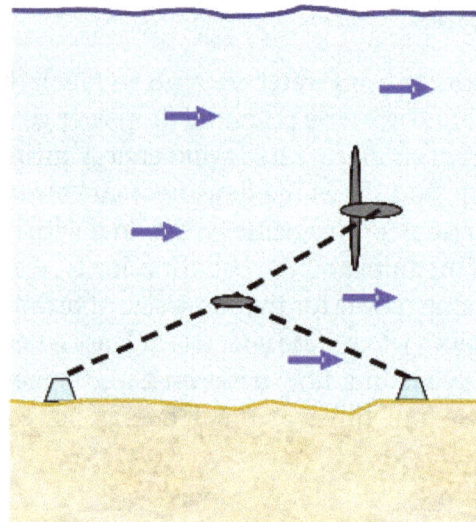

A cable tethered turbine

The AR-1000, a 1MW turbine developed by Atlantis Resources Corporation that was successfully deployed at the EMEC facility during the summer of 2011. The AR series are commercial scale, horizontal axis turbines designed for open ocean deployment. AR turbines feature a single rotor set with fixed pitch blades. The AR turbine is rotated as required with each tidal exchange. This is done in the slack period between tides and held in place for the optimal heading for the next tide. AR turbines are rated at 1MW @ 2.65 m/s of water flow velocity.

The Kvalsund installation is south of Hammerfest, Norway. Although still a prototype, a turbine with a reported capacity of 300 kW was connected to the grid on 13 November 2003.

Seaflow, a 300 kW Periodflow marine current propeller type turbine was installed by Marine Current Turbines off the coast of Lynmouth, Devon, England, in 2003. The 11m diameter turbine generator was fitted to a steel pile which was driven into the seabed. As a prototype, it was connected to a dump load, not to the grid.

In April 2007 Verdant Power began running a prototype project in the East River between Queens and Roosevelt Island in New York City; it was the first major tidal-power project in the United States. The strong currents pose challenges to the design: the blades of the 2006 and 2007 prototypes broke and new reinforced turbines were installed in September 2008.

Following the Seaflow trial, a full-size prototype, called SeaGen, was installed by Marine Current Turbines in Strangford Lough in Northern Ireland in April 2008. The turbine began to generate at full power of just over 1.2 MW in December 2008 and is reported to have fed 150 kW into the grid for the first time on 17 July 2008, and has now contributed more than a gigawatt hour to consumers in Northern Ireland. It is currently the only commercial scale device to have been installed anywhere in the world. SeaGen is made up of two axial flow rotors, each of which drive a generator. The turbines are capable of generating electricity on both the ebb and flood tides because the rotor blades can pitch through 180°.

OpenHydro, an Irish company exploiting the Open-Centre Turbine developed in the U.S., has a prototype being tested at the European Marine Energy Centre (EMEC), in Orkney, Scotland.

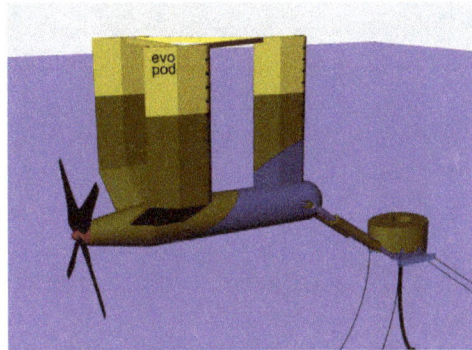

A 3D model of an Evopod tidal turbine

A prototype semi-submerged floating tethered tidal turbine called Evopod has been tested since June 2008 in Strangford Lough, Northern Ireland at 1/10 scale. The UK company developing it is called Ocean Flow Energy Ltd. The advanced hull form maintains optimum heading into the tidal stream and is designed to operate in the peak flow of the water column.

In 2010, Tenax Energy of Australia proposed to put 450 turbines off the coast of Darwin, Australia, in the Clarence Strait. The turbines would feature a rotor section approximately 15 metres in diameter with a slightly larger gravity base. The turbines would operate in deep water well below shipping channels. Each turbine is forecast to produce energy for between 300 and 400 homes.

Tidalstream, a UK-based company, commissioned a scaled-down Triton 3 turbine in the Thames. It can be floated to its site, installed without cranes, jack-ups or divers and then ballasted into operating position. At full scale the Triton 3 in 30-50m deep water has a 3MW capacity, and the Triton 6 in 60-80m water has a capacity of up to 10MW, depending on the flow. Both platforms have man-access capability both in the operating position and in the float-out maintenance position.

Crossflow Turbines

Invented by Georges Darreius in 1923 and patented in 1929, these turbines can be deployed either vertically or horizontally.

The Gorlov turbine is a variant of the Darrieus design featuring a helical design that is in a large scale, commercial pilot in South Korea, starting with a 1MW plant that opened in May 2009 and expanding to 90MW by 2013. Neptune Renewable Energy's Proteus project employs a shrouded vertical axis turbine that can be used to form an array in mainly estuarine conditions.

In April 2008, the Ocean Renewable Power Company, LLC (ORPC) successfully completed testing its proprietary turbine-generator unit (TGU) prototype at ORPC's Cobscook Bay and Western Passage tidal sites near Eastport, Maine. The TGU is the core of the OCGen technology and utilizes advanced design cross-flow (ADCF) turbines to drive a permanent magnet generator located between the turbines and mounted on the same shaft. ORPC has developed TGU designs that can be used for generating power from river, tidal and deep water ocean currents.

Trials in the Strait of Messina, Italy, started in 2001 of the Kobold turbine concept.

Flow Augmented Turbines

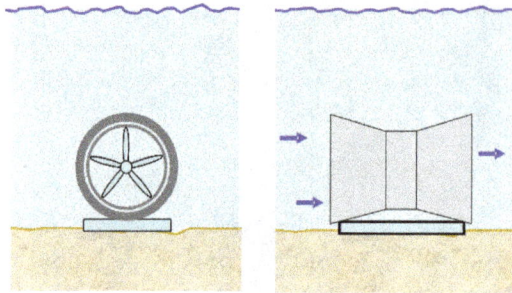

A shrouded turbine

Using flow augmentation measures, for example a duct or shroud, the incident power available to a turbine can be increased. The most common example uses a shroud to increase the flow rate through the turbine, which can be either axial or crossflow.

The Australian company Tidal Energy Pty Ltd undertook successful commercial trials of efficient shrouded tidal turbines on the Gold Coast, Queensland in 2002. Tidal Energy delivered their shrouded turbine in northern Australia where some of the fastest recorded flows (11 m/s, 21 knots) are found. Two small turbines will provide 3.5 MW. Another larger 5 meter diameter turbine, capable of 800 kW in 4 m/s of flow, was planned as a tidal powered desalination showcase near Brisbane Australia.

Oscillating Devices

Oscillating devices do not have a rotating component, instead making use of aerofoil sections which are pushed sideways by the flow. Oscillating stream power extraction was proven with the omni- or bi-directional Wing'd Pump windmill. During 2003 a 150 kW oscillating hydroplane device, the Stingray, was tested off the Scottish coast. The Stingray uses hydrofoils to create oscillation, which allows it to create hydraulic power. This hydraulic power is then used to power a hydraulic motor, which then turns a generator.

Pulse Tidal operate an oscillating hydrofoil device in the Humber estuary. Having secured funding from the EU, they are developing a commercial scale device to be commissioned 2012.

The bioSTREAM tidal power conversion system, uses the biomimicry of swimming species, such as shark, tuna, and mackerel using their highly efficient Thunniform mode propulsion. It is produced by Australian company BioPower Systems.

A 2 kW prototype relying on the use of two oscillating hydrofoils in a tandem configuration has been developed at Laval University and tested successfully near Quebec City, Canada, in 2009. A hydrodynamic efficiency of 40% has been achieved during the field tests.

Venturi Effect

Venturi effect devices use a shroud or duct in order to generate a pressure differential which is used to run a secondary hydraulic circuit which is used to generate power. A device, the Hydro Venturi, is to be tested in San Francisco Bay.

Tidal Kite Turbines

A tidal kite turbine is an underwater kite system or paravane that converts tidal energy into electricity by moving through the tidal stream. An estimated 1% of 2011's global energy requirements could be provided by such devices at scale.

History

Ernst Souczek of Vienna, Austria, on August 6, 1947, filed for a patent US2501696; assignor of one-half to Wolfgang Kmentt, also of Vienna. Their water kite turbine disclosure demonstrated a rich art in water-kite turbines. In similar technology, many others prior to 2006 advanced water-kite and paravane electric generating systems. In 2006, a tidal kite turbine was developed by Swedish company Minesto. They conducted its first sea trial in Strangford Lough in Northern Ireland in the summer of 2011. The test used kites with wingspan of 1.4m. In 2013 the Deep Green pilot plant began operation off Northern Ireland. The plant uses carbon fiber kites with a wingspan of 8m (or 12m). Each kite has a rated power of 120 kilowatts at a tidal flow of 1.3 meters per second.

Design

Minesto's kite has a wingspan of 8–14 metres (26–46 ft). The kite has neutral buoyancy, so doesn't sink as the tide turns from ebb to flow. Each kite is equipped with a gearless turbine to generate which is transmitted by the attachment cable to a transformer and then to the electricity grid. The turbine mouth is protected to protect marine life. The 14-meter version has a rated power of 850 kilowatts at 1.7 meters per second.

Operation

The kite is tethered by a cable to a fixed point. It "flies" through the current carrying a turbine. It moves in a figure-eight loop to increase the speed of the water flowing through the turbine tenfold. Force increases with the cube of velocity, offering the potential to generate 1,000-fold more energy than a stationary generator. That maneuver means the kite can operate in tidal streams that move too slowly to drive earlier tidal devices, such as the SeaGen turbine. The kite was expected to work in flows as low 1–2.5 metres (3 ft 3 in–8 ft 2 in) per second, while first-generation devices need over 2.5s. Each kite will have a capacity to generate between 150 and 800 kW. They can be deployed in waters 50–300 metres (160–980 ft) deep.

Tidal Stream Developers

There are a number of individuals and companies developing tidal energy converters across the world. A database of all know tidal energy developers is kept up-to-date here: Tidal energy developers

Tidal Stream Testing

The world's first marine energy test facility was established in 2003 to kick start the development of the wave and tidal energy industry in the UK. Based in Orkney, Scotland, the European Marine Energy Centre (EMEC) has supported the deployment of more wave and tidal energy devices than at any other single site in the world. EMEC provides a variety of test sites in real sea conditions. It's grid connected tidal test site is located at the Fall of Warness, off the island of Eday, in a narrow channel which concentrates the tide as it flows between the Atlantic Ocean and North Sea. This area has a very strong tidal current, which can travel up to 4 m/s (8 knots) in spring tides. Tidal energy developers currently testing at the site include Alstom (formerly Tidal Generation Ltd), ANDRITZ HYDRO Hammerfest, OpenHydro, Scotrenewables Tidal Power, and Voith.

Commercial Plans

RWE's npower announced that it is in partnership with Marine Current Turbines to build a tidal farm of SeaGen turbines off the coast of Anglesey in Wales, near the Skerries.

"The Skerries project located in Anglesey, Wales, will be one of the first arrays deployed using the Siemens owned Marine Current Turbines SeaGen S tidal turbines. The marine consent for the project was recently awarded, the first tidal array to be consented in Wales. The 10MW array will be fully operational in 2015." - CEO of Siemens Energy Hydro & Ocean Unit Achim Wörner

In November 2007, British company Lunar Energy announced that, in conjunction with E.ON, they would be building the world's first deep-sea tidal energy farm off the coast of Pembrokshire in Wales. It will provide electricity for 5,000 homes. Eight underwater turbines, each 25 metres long and 15 metres high, are to be installed on the sea bottom off St David's peninsula. Construction is due to start in the summer of 2008 and the proposed tidal energy turbines, described as "a wind farm under the sea", should be operational by 2010.

British Columbia Tidal Energy Corp. plans to deploy at least three 1.2 MW turbines in the Campbell River or in the surrounding coastline of British Columbia by 2009.

Alderney Renewable Energy Ltd is planning to use tidal turbines to extract power from the notoriously strong tidal races around Alderney in the Channel Islands. It is estimated that up to 3 GW could be extracted. This would not only supply the island's needs but also leave a considerable surplus for export.

Nova Scotia Power has selected OpenHydro's turbine for a tidal energy demonstration project in the Bay of Fundy, Nova Scotia, Canada and Alderney Renewable Energy Ltd for the supply of tidal turbines in the Channel Islands.

Pulse Tidal are designing a commercial device with seven other companies who are expert in their fields. The consortium was awarded an €8 million EU grant to develop the first device, which will be deployed in 2012 and generate enough power for 1,000 homes.

ScottishPower Renewables are planning to deploy ten 1MW HS1000 devices designed by Hammerfest Strom in the Sound of Islay.

In March 2014, the Federal Energy Regulatory Committee (FERC) approved a pilot license for Snohomish County PUD to install two OpenHydro tidal turbines in Admiralty Inlet, WA. This project is the first grid-connected two-turbine project in the US; installation is planned for the summer of 2015. The OpenHydro tidal turbines that Snohomish County PUD will use are designed to be placed directly into the seafloor at a depth of roughly 200 feet, so that there will be no effect on commercial navigation overhead. The license granted by the FERC also includes plans to protect fish, wildlife, as well as cultural and aesthetic resources, in addition to navigation. Each turbine measures 6 meters in diameter, and will generate up to 300 kW of electricity.

Energy Calculations

Turbine Power

Tidal energy converters can have varying modes of operating and therefore varying power output. If the power coefficient of the device "C_P" is known, the equation below can be used to determine the power output of the hydrodynamic subsystem of the machine. This available power cannot exceed that imposed by the Betz limit on the power coefficient, although this can be circumvented to some degree by placing a turbine in a shroud or duct. This works, in essence, by forcing water which would not have flowed through the turbine through the rotor disk. In these situations it is the frontal area of the duct, rather than the turbine, which is used in calculating the power coefficient and therefore the Betz limit still applies to the device as a whole.

The energy available from these kinetic systems can be expressed as:

$$P = \frac{\rho A V^3}{2} C_P$$

where:

C_P = the turbine power coefficient

P = the power generated (in watts)

ρ = the density of the water (seawater is 1027 kg/m³)

A = the sweep area of the turbine (in m²)

V = the velocity of the flow

Relative to an open turbine in free stream, ducted turbines are capable of as much as 3 to 4 times the power of the same turbine rotor in open flow.

Resource Assessment

While initial assessments of the available energy in a channel have focus on calculations using the kinetic energy flux model, the limitations of tidal power generation are significantly more complicated. For example, the maximum physical possible energy extraction from a strait connecting two large basins is given to within 10% by:

$$P = 0.22 \rho g \Delta H_{max} Q_{max}$$

where

ρ = the density of the water (seawater is 1027 kg/m³)

g = gravitational acceleration (9.80665 m/s²)

ΔH_{max} = maximum differential water surface elevation across the channel

Q_{max} = maximum volumetric flow rate though the channel.

Potential Sites

As with wind power, selection of location is critical for the tidal turbine. Tidal stream systems need to be located in areas with fast currents where natural flows are concentrated between obstructions, for example at the entrances to bays and rivers, around rocky points, headlands, or between islands or other land masses. The following potential sites are under serious consideration:

- Pembrokeshire in Wales

- River Severn between Wales and England

- Cook Strait in New Zealand

- Kaipara Harbour in New Zealand

- Bay of Fundy in Canada.

- East River in the United States

- Golden Gate in the San Francisco Bay

- Piscataqua River in New Hampshire

- The Race of Alderney and The Swinge in the Channel Islands

- The Sound of Islay, between Islay and Jura in Scotland

- Pentland Firth between Caithness and the Orkney Islands, Scotland

- Humboldt County, California in the United States

- Columbia River, Oregon in the United States

- Plaquemines Parish, Louisiana in the Southern United States

- Isle of Wight, England

- Teddington and Ham Hydro at Teddington on the River Thames in the London suburbs, England

Modern advances in turbine technology may eventually see large amounts of power generated from the ocean, especially tidal currents using the tidal stream designs but also from the major thermal current systems such as the Gulf Stream, which is covered by the more general term marine current power. Tidal stream turbines may be arrayed in high-velocity areas where natural tidal current flows are concentrated such as the west and east coasts of Canada, the Strait of Gibraltar, the Bosporus, and numerous sites in Southeast Asia and Australia. Such flows occur almost anywhere where there are entrances to bays and rivers, or between land masses where water currents are concentrated.

Environmental Impacts

The main environmental concern with tidal energy is associated with blade strike and entanglement of marine organisms as high speed water increases the risk of organisms being pushed near or through these devices. As with all offshore renewable energies, there is also a concern about how the creation of EMF and acoustic outputs may affect marine organisms. It should be noted that because these devices are in the water, the acoustic output can be greater than those created with offshore wind energy. Depending on the frequency and amplitude of sound generated by the tidal energy devices, this acoustic output can have varying effects on marine mammals (particularly those who echolocate to communicate and navigate in the marine environment such as dolphins and whales). Tidal energy removal can also cause environmental concerns such as degrading farfield water quality and disrupting sediment processes. Depending on the size of the project, these effects can range from small traces of sediment build up near the tidal device to severely affecting nearshore ecosystems and processes.

One study of the Roosevelt Island Tidal Energy (RITE, Verdant Power) project in the East River (New York City), utilized 24 split beam hydroacoustic sensors (scientific echosounder) to detect and track the movement of fish both upstream and downstream of each of six turbines. The results suggested (1) very few fish using this portion of the river, (2) those fish which did use this area were not using the portion of the river which would subject them to blade strikes, and (3) no evidence of fish traveling through blade areas.

Work is currently being conducted by the Northwest National Marine Renewable Energy Center (NNMREC)to explore and establish tools and protocols for assessment of physical and biological conditions and monitor environmental changes associated with tidal energy development.

Windbelt

The Windbelt is a wind power harvesting device invented by Shawn Frayn in 2004 for converting wind power to electricity. It consists of a flexible polymer ribbon stretched between supports transverse to the wind direction, with magnets glued to it. When the wind blows across it, the ribbon vibrates due to aeroelastic flutter, similar to the action of an aeolian harp. The vibrating movement of the magnets induces current in nearby pickup coils by electromagnetic induction.

One prototype has powered two LEDs, a radio, and a clock (separately) using wind generated from a household fan. The cost of the materials was well under US$10. $2–$5 for 40 mW is a cost of $50–$125 per watt.

There are three sizes in development:

The microBelt, a 12 cm version. This could be put into production in around six months. Its expected to produce 1 milliwatt average. To charge a pair of ideal rechargeable AA cells (2.5Ah 1.2v) this would take 6000 hours, or 250 days.

The Windcell, a 1-metre version that could be used to power meshed WiFi repeaters, charge cellphones, or run LED lights. This could go into production within 18 to 24 months. It is hoped that a square metre panel at 6 m/s average windspeed can generate 10 W average.

an experimental 10-metre model that has no production date.

The Windbelt's inventor, Shawn Frayn, was a winner of the 2007 Breakthrough Award from the publishers of the magazine, *Popular Mechanics*. He is trying to make the Windbelt cheaper.

The inventor's claims that the device is 10 - 30 times more efficient than small wind turbines have been refuted by tests. The microWindbelt could generate 0.2 mW at a wind speed of 3.5 m/s and 5 mW at 7.5 m/s, which represent efficiencies (ηC_p) of 0.21% and 0.53% respectively. Wind turbines typically have efficiencies of 1% to 10%. Since the Windbelt a number of other "flutter" wind harvester devices have been designed, but like the Windbelt almost all have efficiencies below turbine machines.

References

- Ahmad Y Hassan, Donald Routledge Hill (1986). Islamic Technology: An illustrated history, p. 54. Cambridge University Press. ISBN 0-521-42239-6.

- Morthorst, Poul Erik; Redlinger, Robert Y.; Andersen, Per (2002). Wind energy in the 21st century: economics, policy, technology and the changing electricity industry. Houndmills, Basingstoke, Hampshire: Palgrave/UNEP. ISBN 0-333-79248-3.

- Bent Sørensen (2004). Renewable Energy: Its Physics, Engineering, Use, Environmental Impacts, Economy, and Planning Aspects. Academic Press. pp. 556–. ISBN 978-0-12-656153-1.

- Gevorkian, Peter (2007). Sustainable energy systems engineering: the complete green building design resource. McGraw Hill Professional. ISBN 978-0-07-147359-0.

- Pearce, J.; Lau, A. (2002). "Net Energy Analysis for Sustainable Energy Production from Silicon Based Solar Cells". Solar Energy (PDF). p. 181. doi:10.1115/SED2002-1051. ISBN 0-7918-1689-3.

- Kim, D.S.; et al. (18 May 2003). "String ribbon silicon solar cells with 17.8% efficiency" (PDF). Proceedings of 3rd World Conference on Photovoltaic Energy Conversion, 2003. 2: 1293–1296. ISBN 4-9901816-0-3.

- Morten Lund (30 May 2016). "Dansk firma sætter prisbelønnet selvhejsende kran i serieproduktion". Ingeniøren. Retrieved 30 May 2016.

- Tom Gray (11 March 2013). "Fact check: About those 'abandoned' turbines ...". American Wind Energy Association. Retrieved 30 May 2016.

- Janz, Stefan; Reber, Stefan (14 September 2015). "20% Efficient Solar Cell on EpiWafer". Fraunhofer ISE. Retrieved October 15, 2015.

- "Special Operations Summit Little Creek – Wind Power: "Small and Micro Wind"...A Position of Power I". Tactical Defense Media's DoD Power, Energy & Propulsion Q2 2012, tacticaldefensemedia.com. Retrieved 2015-10-20.

- Wittrup, Sanne. "Power from Vestas' giant turbine" (in Danish. English translation). Ingeniøren, 28 January 2014. Retrieved 28 January 2014.

- "Technology Roadmap: Solar Photovoltaic Energy" (PDF). IEA. 2014. Archived from the original on 7 October 2014. Retrieved 7 October 2014.

- "French-German collaborators claim solar cell efficiency world record". EE Times Europe. 2 December 2014. Retrieved 3 December 2014.

- "Photovoltaics Report" (PDF). Fraunhofer ISE. 28 July 2014. Archived from the original on 31 August 2014. Retrieved 31 August 2014.

- PV Status Report 2013 | Renewable Energy Mapping and Monitoring in Europe and Africa (REMEA). Iet.jrc.ec.europa.eu (11 April 2014). Retrieved on 20 April 2014.

Semiconductor Devices: An Overview

Semiconductor devices utilize semiconductor materials and their conductivity can be regulated or manipulated by doping or the application of an electromagnetic field, light or even heat. This chapter explores semiconductor devices like power semiconductor device, diode, zener diode, tunnel diode, thyristor and transistor. The content focuses on the operation, construction and uses of these devices. This chapter elucidates the crucial theories and applications of semiconductor devices.

Semiconductor Device

Semiconductor devices are electronic components that exploit the electronic properties of semiconductor materials, principally silicon, germanium, and gallium arsenide, as well as organic semiconductors. Semiconductor devices have replaced thermionic devices (vacuum tubes) in most applications. They use electronic conduction in the solid state as opposed to the gaseous state or thermionic emission in a high vacuum.

Semiconductor devices are manufactured both as single discrete devices and as *integrated circuits* (ICs), which consist of a number—from a few (as low as two) to billions—of devices manufactured and interconnected on a single semiconductor substrate, or wafer.

Semiconductor materials are useful because their behavior can be easily manipulated by the addition of impurities, known as doping. Semiconductor conductivity can be controlled by the introduction of an electric or magnetic field, by exposure to light or heat, or by the mechanical deformation of a doped monocrystalline grid; thus, semiconductors can make excellent sensors. Current conduction in a semiconductor occurs via mobile or "free" *electrons* and *holes*, collectively known as *charge carriers*. Doping a semiconductor such as silicon with a small amount of impurity atoms, such as phosphorus or boron, greatly increases the number of free electrons or holes within the semiconductor. When a doped semiconductor contains excess holes it is called "p-type", and when it contains excess free electrons it is known as "n-type", where p (positive for holes) or n (negative for electrons) is the sign of the charge of the majority mobile charge carriers. The semiconductor material used in devices is doped under highly controlled conditions in a fabrication facility, or *fab*, to control precisely the location and concentration of p- and n-type dopants. The junctions which form where n-type and p-type semiconductors join together are called p–n junctions.

A semiconductor diode is a device typically made from a single p–n junction. At the junction of a p-type and an n-type semiconductor there forms a depletion region where current conduction is inhibited by the lack of mobile charge carriers. When the device is *forward biased* (connected with the p-side at higher electric potential than the n-side), this depletion region is diminished, allowing for significant conduction, while only very small current can be achieved when the diode is *reverse biased* and thus the depletion region expanded.

Exposing a semiconductor to light can generate electron–hole pairs, which increases the number of free carriers and thereby the conductivity. Diodes optimized to take advantage of this phenomenon are known as *photodiodes*. Compound semiconductor diodes can also be used to generate light, as in light-emitting diodes and laser diodes.

Transistor

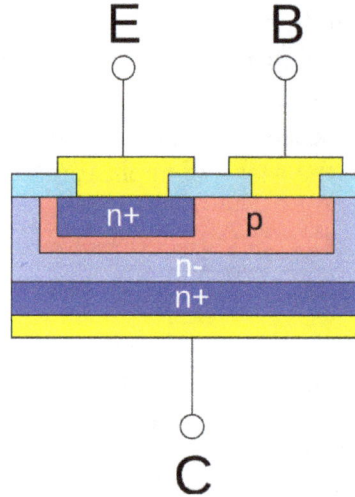

An n–p–n bipolar junction transistor structure

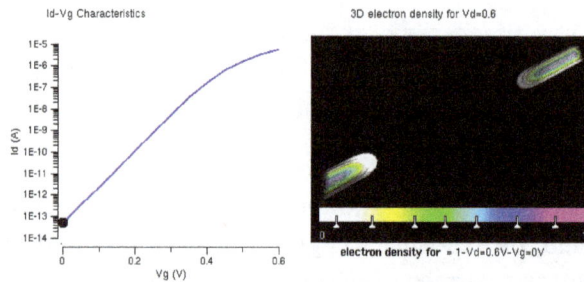

Operation of a MOSFET and its Id-Vg curve. At first, when no gate voltage is applied. There is no inversion electron in the channel, the device is OFF. As gate voltage increase, inversion electron density in the channel increase, current increase, the device turns on.

Bipolar Junction Transistor

Bipolar junction transistors are formed from two p–n junctions, in either n–p–n or p–n–p configuration. The middle, or *base*, region between the junctions is typically very narrow. The other regions, and their associated terminals, are known as the *emitter* and the *collector*. A small current injected through the junction between the base and the emitter changes the properties of the base-collector junction so that it can conduct current even though it is reverse biased. This creates a much larger current between the collector and emitter, controlled by the base-emitter current.

Field-Effect Transistor

Another type of transistor, the field-effect transistor, operates on the principle that semiconductor conductivity can be increased or decreased by the presence of an electric field. An electric field can increase the number of free electrons and holes in a semiconductor, thereby changing its conductivity. The field may be applied by a reverse-biased p–n junction, forming a *junction field-effect transistor* (JFET) or by an electrode insulated from the bulk material by an oxide layer, forming a *metal–oxide–semiconductor field-effect transistor* (MOSFET).

The MOSFET, a solid-state device, is the most used semiconductor device today. The *gate* electrode is charged to produce an electric field that controls the conductivity of a "channel" between two terminals, called the *source* and *drain*. Depending on the type of carrier in the channel, the device may be an *n-channel* (for electrons) or a *p-channel* (for holes) MOSFET. Although the MOSFET is named in part for its "metal" gate, in modern devices polysilicon is typically used instead.

Semiconductor Device Materials

By far, silicon (Si) is the most widely used material in semiconductor devices. Its combination of low raw material cost, relatively simple processing, and a useful temperature range makes it currently the best compromise among the various competing materials. Silicon used in semiconductor device manufacturing is currently fabricated into boules that are large enough in diameter to allow the production of 300 mm (12 in.) wafers.

Germanium (Ge) was a widely used early semiconductor material but its thermal sensitivity makes it less useful than silicon. Today, germanium is often alloyed with silicon for use in very-high-speed SiGe devices; IBM is a major producer of such devices.

Gallium arsenide (GaAs) is also widely used in high-speed devices but so far, it has been difficult to form large-diameter boules of this material, limiting the wafer diameter to sizes significantly smaller than silicon wafers thus making mass production of GaAs devices significantly more expensive than silicon.

Other less common materials are also in use or under investigation.

Silicon carbide (SiC) has found some application as the raw material for blue light-emitting diodes (LEDs) and is being investigated for use in semiconductor devices that could withstand very high operating temperatures and environments with the presence of significant levels of ionizing radiation. IMPATT diodes have also been fabricated from SiC.

Various indium compounds (indium arsenide, indium antimonide, and indium phosphide) are also being used in LEDs and solid state laser diodes. Selenium sulfide is being studied in the manufacture of photovoltaic solar cells.

The most common use for organic semiconductors is Organic light-emitting diodes.

List of Common Semiconductor Devices

Two-terminal devices:

- DIAC

- Diode (rectifier diode)
- Gunn diode
- IMPATT diode
- Laser diode
- Light-emitting diode (LED)
- Photocell
- Phototransistor
- PIN diode
- Schottky diode
- Solar cell
- Transient-voltage-suppression diode
- Tunnel diode
- VCSEL
- Zener diode

Three-terminal devices:

- Bipolar transistor
- Darlington transistor
- Field-effect transistor
- Insulated-gate bipolar transistor (IGBT)
- Silicon-controlled rectifier
- Thyristor
- TRIAC
- Unijunction transistor

Four-terminal devices:

- Hall effect sensor (magnetic field sensor)
- Photocoupler (Optocoupler)

Semiconductor Device Applications

All transistor types can be used as the building blocks of logic gates, which are fundamental in the design of digital circuits. In digital circuits like microprocessors, transistors act as on-off switches; in the MOSFET, for instance, the voltage applied to the gate determines whether the switch is on or off.

Transistors used for analog circuits do not act as on-off switches; rather, they respond to a continuous range of inputs with a continuous range of outputs. Common analog circuits include amplifiers and oscillators.

Circuits that interface or translate between digital circuits and analog circuits are known as mixed-signal circuits.

Power semiconductor devices are discrete devices or integrated circuits intended for high current or high voltage applications. Power integrated circuits combine IC technology with power semiconductor technology, these are sometimes referred to as "smart" power devices. Several companies specialize in manufacturing power semiconductors.

Component Identifiers

The type designators of semiconductor devices are often manufacturer specific. Nevertheless, there have been attempts at creating standards for type codes, and a subset of devices follow those. For discrete devices, for example, there are three standards: JEDEC JESD370B in United States, Pro Electron in Europe and Japanese Industrial Standards (JIS) in Japan.

History of Semiconductor Device Development

Cat's-Whisker Detector

Semiconductors had been used in the electronics field for some time before the invention of the transistor. Around the turn of the 20th century they were quite common as detectors in radios, used in a device called a "cat's whisker" developed by Jagadish Chandra Bose and others. These detectors were somewhat troublesome, however, requiring the operator to move a small tungsten filament (the whisker) around the surface of a galena (lead sulfide) or carborundum (silicon carbide) crystal until it suddenly started working. Then, over a period of a few hours or days, the cat's whisker would slowly stop working and the process would have to be repeated. At the time their operation was completely mysterious. After the introduction of the more reliable and amplified vacuum tube based radios, the cat's whisker systems quickly disappeared. The "cat's whisker" is a primitive example of a special type of diode still popular today, called a Schottky diode.

Metal Rectifier

Another early type of semiconductor device is the metal rectifier in which the semiconductor is copper oxide or selenium. Westinghouse Electric (1886) was a major manufacturer of these rectifiers.

World War II

During World War II, radar research quickly pushed radar receivers to operate at ever higher frequencies and the traditional tube based radio receivers no longer worked well. The introduction of the cavity magnetron from Britain to the United States in 1940 during the Tizard Mission resulted in a pressing need for a practical high-frequency amplifier.

On a whim, Russell Ohl of Bell Laboratories decided to try a cat's whisker. By this point they had not been in use for a number of years, and no one at the labs had one. After hunting one down at a used radio store in Manhattan, he found that it worked much better than tube-based systems.

Ohl investigated why the cat's whisker functioned so well. He spent most of 1939 trying to grow more pure versions of the crystals. He soon found that with higher quality crystals their finicky behaviour went away, but so did their ability to operate as a radio detector. One day he found one of his purest crystals nevertheless worked well, and interestingly, it had a clearly visible crack near the middle. However as he moved about the room trying to test it, the detector would mysteriously work, and then stop again. After some study he found that the behaviour was controlled by the light in the room—more light caused more conductance in the crystal. He invited several other people to see this crystal, and Walter Brattain immediately realized there was some sort of junction at the crack.

Further research cleared up the remaining mystery. The crystal had cracked because either side contained very slightly different amounts of the impurities Ohl could not remove—about 0.2%. One side of the crystal had impurities that added extra electrons (the carriers of electric current) and made it a "conductor". The other had impurities that wanted to bind to these electrons, making it (what he called) an "insulator". Because the two parts of the crystal were in contact with each other, the electrons could be pushed out of the conductive side which had extra electrons (soon to be known as the *emitter*) and replaced by new ones being provided (from a battery, for instance) where they would flow into the insulating portion and be collected by the whisker filament (named the *collector*). However, when the voltage was reversed the electrons being pushed into the collector would quickly fill up the "holes" (the electron-needy impurities), and conduction would stop almost instantly. This junction of the two crystals (or parts of one crystal) created a solid-state diode, and the concept soon became known as semiconduction. The mechanism of action when the diode is off has to do with the separation of charge carriers around the junction. This is called a "depletion region".

Development of The Diode

Armed with the knowledge of how these new diodes worked, a vigorous effort began to learn how to build them on demand. Teams at Purdue University, Bell Labs, MIT, and the University of Chicago all joined forces to build better crystals. Within a year germanium production had been perfected to the point where military-grade diodes were being used in most radar sets.

Development of The Transistor

After the war, William Shockley decided to attempt the building of a triode-like semiconductor device. He secured funding and lab space, and went to work on the problem with Brattain and John Bardeen.

The key to the development of the transistor was the further understanding of the process of the electron mobility in a semiconductor. It was realized that if there were some way to control the flow of the electrons from the emitter to the collector of this newly discovered diode, an amplifier could be built. For instance, if contacts are placed on both sides of a single type of crystal, current will not flow between them through the crystal. However if a third contact could then "inject" electrons or holes into the material, current would flow.

Actually doing this appeared to be very difficult. If the crystal were of any reasonable size, the number of electrons (or holes) required to be injected would have to be very large, making it less than useful as an amplifier because it would require a large injection current to start with. That said, the whole idea of the crystal diode was that the crystal itself could provide the electrons over a very small distance, the depletion region. The key appeared to be to place the input and output contacts very close together on the surface of the crystal on either side of this region.

Brattain started working on building such a device, and tantalizing hints of amplification continued to appear as the team worked on the problem. Sometimes the system would work but then stop working unexpectedly. In one instance a non-working system started working when placed in water. Ohl and Brattain eventually developed a new branch of quantum mechanics, which became known as surface physics, to account for the behaviour. The electrons in any one piece of the crystal would migrate about due to nearby charges. Electrons in the emitters, or the "holes" in the collectors, would cluster at the surface of the crystal where they could find their opposite charge "floating around" in the air (or water). Yet they could be pushed away from the surface with the application of a small amount of charge from any other location on the crystal. Instead of needing a large supply of injected electrons, a very small number in the right place on the crystal would accomplish the same thing.

Their understanding solved the problem of needing a very small control area to some degree. Instead of needing two separate semiconductors connected by a common, but tiny, region, a single larger surface would serve. The electron-emitting and collecting leads would both be placed very close together on the top, with the control lead placed on the base of the crystal. When current flowed through this "base" lead, the electrons or holes would be pushed out, across the block of semiconductor, and collect on the far surface. As long as the emitter and collector were very close together, this should allow enough electrons or holes between them to allow conduction to start.

The First Transistor

The Bell team made many attempts to build such a system with various tools, but generally failed. Setups where the contacts were close enough were invariably as fragile as the original cat's whisker detectors had been, and would work briefly, if at all. Eventually they had a practical breakthrough. A piece of gold foil was glued to the edge of a plastic wedge, and then the foil was sliced with a razor at the tip of the triangle. The result was two very closely spaced contacts of gold. When the wedge was pushed down onto the surface of a crystal and voltage applied to the other side (on the base of the crystal), current started to flow from one contact to the other as the base voltage pushed the electrons away from the base towards the other side near the contacts. The point-contact transistor had been invented.

While the device was constructed a week earlier, Brattain's notes describe the first demonstration to higher-ups at Bell Labs on the afternoon of 23 December 1947, often given as the birthdate of the transistor. what is now known as the "p–n–p point-contact germanium transistor" operated as a speech amplifier with a power gain of 18 in that trial. John Bardeen, Walter Houser Brattain, and William Bradford Shockley were awarded the 1956 Nobel Prize in physics for their work.

A stylized replica of the first transistor

Origin of The Term "Transistor"

Bell Telephone Laboratories needed a generic name for their new invention: "Semiconductor Triode", "Solid Triode", "Surface States Triode" [sic], "Crystal Triode" and "Iotatron" were all considered, but "transistor", coined by John R. Pierce, won an internal ballot. The rationale for the name is described in the following extract from the company's Technical Memoranda (May 28, 1948) calling for votes:

Transistor. This is an abbreviated combination of the words "transconductance" or "transfer", and "varistor". The device logically belongs in the varistor family, and has the transconductance or transfer impedance of a device having gain, so that this combination is descriptive.

Improvements in Transistor Design

Shockley was upset about the device being credited to Brattain and Bardeen, who he felt had built it "behind his back" to take the glory. Matters became worse when Bell Labs lawyers found that some of Shockley's own writings on the transistor were close enough to those of an earlier 1925 patent by Julius Edgar Lilienfeld that they thought it best that his name be left off the patent application.

Shockley was incensed, and decided to demonstrate who was the real brains of the operation. A few months later he invented an entirely new, considerably more robust, type of transistor with a layer or 'sandwich' structure. This structure went on to be used for the vast majority of all transistors into the 1960s, and evolved into the bipolar junction transistor.

With the fragility problems solved, a remaining problem was purity. Making germanium of the required purity was proving to be a serious problem, and limited the yield of transistors that actually worked from a given batch of material. Germanium's sensitivity to temperature also limited its usefulness. Scientists theorized that silicon would be easier to fabricate, but few investigated this possibility. Gordon K. Teal was the first to develop a working silicon transistor, and his company, the nascent Texas Instruments, profited from its technological edge. From the late 1960s most transistors were silicon-based. Within a few years transistor-based products, most notably easily portable radios, were appearing on the market.

A major improvement in manufacturing yield came when a chemist advised the companies fabricating semiconductors to use distilled rather than tap water: calcium ions present in tap water were the cause of the poor yields. "Zone melting", a technique using a band of molten material moving through the crystal, further increased crystal purity.

Power Semiconductor Device

A power semiconductor device is a semiconductor device used as a switch or rectifier in power electronics; a switch-mode power supply is an example. Such a device is also called a power device or, when used in an integrated circuit, a power IC.

A power semiconductor device is usually used in "commutation mode" (i.e., it is either on or off), and therefore has a design optimized for such usage; it should usually not be used in linear operation.

History

The first power semiconductor device appeared in 1952 with the introduction of the power diode by R.N. Hall. It was made of germanium and had a reverse voltage blocking capability of 200 V and a current rating of 35 A.

The thyristor appeared in 1957. It is able to withstand very high reverse breakdown voltage and is also capable of carrying high current. However, one disadvantage of the thyristor in switching circuits is that once it becomes 'latched-on' in the conducting state; it cannot be turned off by external control, as the thyristor turn-off is passive, i.e., the power must be disconnected from the device. Thyristors which could be turned off, called gate turn-off thyristors (GTO), were introduced in 1960. These overcome some limitations of the ordinary thyristor, because they can be turned on or off with an applied signal.

The first bipolar transistor device with substantial power handling capabilities was introduced in the 1948 by William Shockley.

Due to improvements in the MOSFET technology (metal oxide semiconductor technology, initially developed to produce integrated circuits), the power MOSFET became available in the late 1970s. International Rectifier introduced a 25 A, 400 V power MOSFET in 1978. This device allows operation at higher frequencies than a bipolar transistor, but is limited to low voltage applications.

The Insulated-gate bipolar transistor (IGBT) was developed in the 1980s, and became widely available in the 1990s. This component has the power handling capability of the bipolar transistor and the advantages of the isolated gate drive of the power MOSFET.

Common Devices

Some common power devices are the power diode, thyristor, power MOSFET, and IGBT. The power diode and power MOSFET operate on similar principles to their low-power counterparts, but are able to carry a larger amount of current and are typically able to support a larger reverse-bias voltage in the *off-state*.

Structural changes are often made in a power device in order to accommodate the higher current density, higher power dissipation, and/or higher reverse breakdown voltage. The vast majority of the discrete (i.e., non-integrated) power devices are built using a vertical structure, whereas small-signal devices employ a lateral structure. With the vertical structure, the current rating of the device is proportional to its area, and the voltage blocking capability is achieved in the height of the die. With this structure, one of the connections of the device is located on the bottom of the semiconductor die.

Classifications

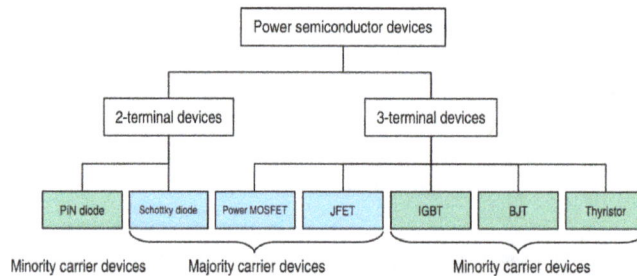

Fig. 1: The power devices family, showing the principal power switches.

A power device may be classified as one of the following main categories:

- A two-terminal device (e.g., a diode), whose state is completely dependent on the external power circuit to which it is connected.

- A three-terminal device (e.g., a triode), whose state is dependent on not only its external power circuit, but also the signal on its driving terminal (this terminal is known as the *gate* or *base*).

Another classification is less obvious, but has a strong influence on device performance:

- A *majority carrier device* (e.g., a Schottky diode, a MOSFET, etc.); this uses only one type of charge carriers.

- A *minority carrier device* (e.g., a thyristor, a bipolar transistor, an IGBT, etc.); this uses both majority and minority carriers (i.e., electrons and electron holes).

A majority carrier device is faster, but the charge injection of minority carrier devices allows for better on-state performance.

Diodes

An ideal diode should have the following characteristics:

- When *forward-biased*, the voltage across the end terminals of the diode should be zero, whatever the current that flows through it (on-state).

- When *reverse-biased*, the leakage current should be zero, whatever the voltage (off-state).

- The transition (or commutation) between the on-state and the off-state should be instantaneous.

In reality, the design of a diode is a trade-off between performance in on-state, off-state, and commutation. Indeed, the same area of the device must sustain the blocking voltage in the off-state and allow current flow in the on-state; as the requirements for the two states are completely opposite, a diode has to be either optimised for one of them, or time must be allowed to switch from one state to the other (i.e., the commutation speed must be reduced).

These trade-offs are the same for all power devices; for instance, a Schottky diode has excellent switching speed and on-state performance, but a high level of leakage current in the off-state. On the other hand, a PIN diode is commercially available in different commutation speeds (what are called "fast" and "ultrafast" rectifiers), but any increase in speed is necessarily associated with a lower performance in the on-state.

Switches

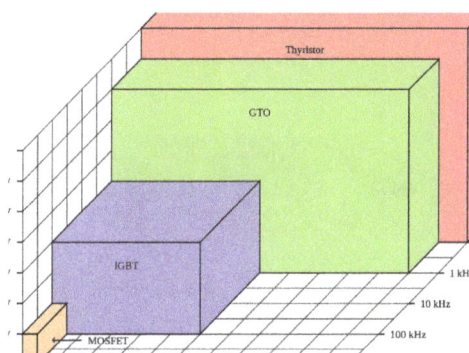

Fig.2 : Current/Voltage/switching frequency domains of the main power electronics switches.

The trade-offs between voltage, current, and frequency ratings also exist for a switch. In fact, any power semiconductor relies on a PIN diode structure in order to sustain voltage; this can be seen in figure 2. The power MOSFET has the advantages of a majority carrier device, so it can achieve a very high operating frequency, but it cannot be used with high voltages; as it is a physical limit, no improvement is expected in the design of a silicon MOSFET concerning its maximum voltage ratings. However, its excellent performance in low voltage applications make it the device of choice (actually the only choice, currently) for applications with voltages below 200 V. By placing several devices in parallel, it is possible to increase the current rating of a switch. The MOSFET is particularly suited to this configuration, because its positive thermal coefficient of resistance tends to result in a balance of current between the individual devices.

The IGBT is a recent component, so its performance improves regularly as technology evolves. It has already completely replaced the bipolar transistor in power applications; a power module is available in which several IGBT devices are connected in parallel, making it attractive for power levels up to several megawatts, which pushes further the limit at which thyristors and GTOs become the only option. Basically, an IGBT is a bipolar transistor driven by a power MOSFET; it has the advantages of being a minority carrier device (good performance in the on-state, even for high voltage devices), with the high input impedance of a MOSFET (it can be driven on or off with a very low amount of power).

The major limitation of the IGBT for low voltage applications is the high voltage drop it exhibits in the on-state (2-to-4 V). Compared to the MOSFET, the operating frequency of the IGBT is relatively low (usually not higher than 50 kHz), mainly because of a problem during turn-off known as *current-tail*: The slow decay of the conduction current during turn-off results from a slow recombination of a large number of carriers that flood the thick 'drift' region of the IGBT during conduction. The net result is that the turn-off switching loss of an IGBT is considerably higher than its turn-on loss. Generally, in datasheets, turn-off energy is mentioned as a measured parameter; that number has to be multiplied with the switching frequency of the intended application in order to estimate the turn-off loss.

At very high power levels, a thyristor-based device (e.g., a SCR, a GTO, a MCT, etc.) is still the only choice. This device can be turned on by a pulse provided by a driving circuit, but cannot be turned off by removing the pulse. A thyristor turns off as soon as no more current flows through it; this happens automatically in an alternating current system on each cycle, or requires a circuit with the means to divert current around the device. Both MCTs and GTOs have been developed to overcome this limitation, and are widely used in power distribution applications.

Parameters

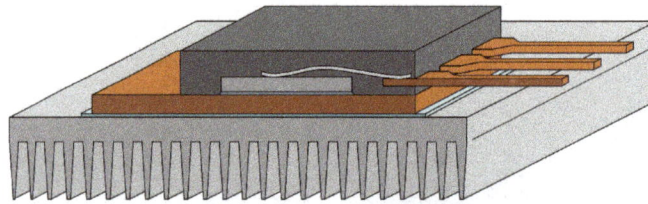

A power device is usually attached to a heatsink to remove the heat caused by operation losses.

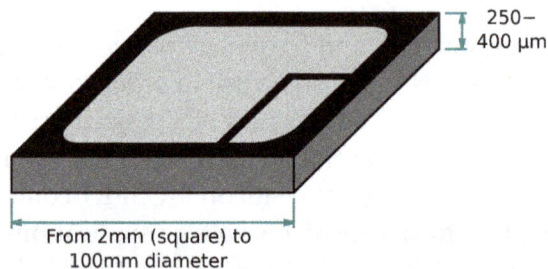

250–400 μm

From 2mm (square) to 100mm diameter

The power semiconductor die of a three-terminal device (IGBT, MOSFET or BJT). Two contacts are on top of the die, the remaining one is on the back.

- Breakdown voltage: Often, there is a trade-off between breakdown voltage rating and on-resistance, because increasing the breakdown voltage by incorporating a thicker and lower doped drift region leads to a higher on-resistance.

- On-resistance: A higher current rating lowers the on-resistance due to greater numbers of parallel cells. This increases overall capacitance and slows down the speed.

- Rise and fall times: The amount of time it takes to switch between the on-state and the off-state.

- Safe-operating area: This is a thermal dissipation and "latch-up" consideration.

- Thermal resistance: This is an often ignored but extremely important parameter from the point of view of practical design; a semiconductor does not perform well at elevated temperature, and yet due to large current conduction, a power semiconductor device invariably heats up. Therefore, such a devices needs to be cooled by removing that heat continuously; packaging and heatsink technology provide a means for removing heat from a semiconductor device by conducting it to the external environment. Generally, a large current device has a large die and packaging surface areas and lower thermal resistance.

Research and Development

Packaging

The role of packaging is to:

- connect a die to the external circuit.

- provide a way to remove the heat generated by the device.

- protect the die from the external environment (moisture, dust, etc.).

Many of the reliability issues of a power device are either related to excessive temperature or fatigue due to thermal cycling. Research is currently carried out on the following topics:

- Cooling performance.

- Resistance to thermal cycling by closely matching the Coefficient of thermal expansion of the packaging to that of the silicon.

- The maximum operating temperature of the packaging material.

Research is also ongoing on electrical issues such as reducing the parasitic inductance of packaging; this inductance limits the operating frequency, because it generates losses during commutation.

A low-voltage MOSFET is also limited by the parasitic resistance of its package, as its intrinsic on-state resistance can be as low as one or two milliohms.

Some of the most common type of power semiconductor packages include the TO-220, TO-247, TO-262, TO-3, D²Pak, etc.

Improvement of Structures

The IGBT design is still under development and can be expected to provide increases in operating voltages. At the high-power end of the range, the MOS-controlled thyristor is a promising device. Achieving a major improvement over the conventional MOSFET structure by employing the super junction charge-balance principle: essentially, it allows the thick drift region of a power MOSFET to be heavily doped, thereby reducing the electrical resistance to electron flow without compromising the breakdown voltage. This is juxtaposed with a region that is similarly doped with the opposite carrier polarity (*holes*); these two similar, but oppositely doped regions effectively cancel out their mobile charge and develop a 'depleted region' that supports the high voltage during the off-state. On the other hand, during the on-state, the higher doping of the drift region allows for the easy flow of carriers, thereby reducing on-resistance. Commercial devices, based on this super junction principle, have been developed by companies like Infineon (CoolMOS™ products) and International Rectifier (IR).

Wide Band-Gap Semiconductors

The major breakthrough in power semiconductor devices is expected from the replacement of silicon by a wide band-gap semiconductor. At the moment, silicon carbide (SiC) is considered to be the most promising. A SiC Schottky diode with a breakdown voltage of 1200 V is commercially available, as is a 1200 V JFET. As both are majority carrier devices, they can operate at high speed. A bipolar device is being developed for higher voltages (up to 20 kV). Among its advantages, silicon carbide can operate at a higher temperature (up to 400 °C) and has a lower thermal resistance than silicon, allowing for better cooling.

Diode

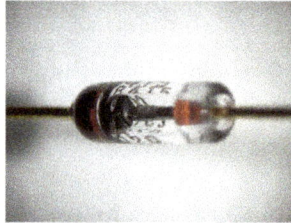

Closeup of a diode, showing the square-shaped semiconductor crystal *(black object on left)*.

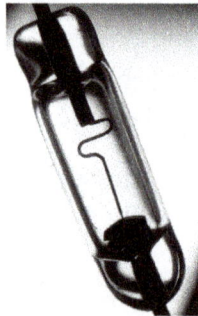

extreme macro photo of a Chinese diode of the seventies

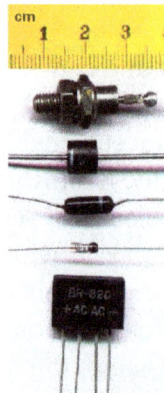

Various semiconductor diodes. Bottom: A bridge rectifier. In most diodes, a white or black painted band identifies the cathode into which electrons will flow when the diode is conducting. Electron flow is the reverse of conventional current flow.

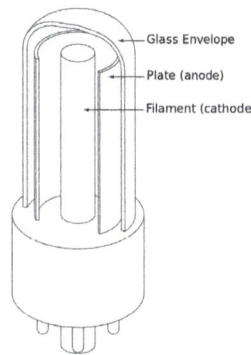

- Glass Envelope
- Plate (anode)
- Filament (cathode

Structure of a vacuum tube diode. The filament may be bare, or more commonly (as shown here), embedded within and insulated from an enclosing cathode.

In electronics, a diode is a two-terminal electronic component that conducts primarily in one direction (asymmetric conductance); it has low (ideally zero) resistance to the flow of current in one direction, and high (ideally infinite) resistance in the other. A semiconductor diode, the most common type today, is a crystalline piece of semiconductor material with a p–n junction connected to two electrical terminals. A vacuum tube diode has two electrodes, a plate (anode) and a heated cathode. Semiconductor diodes were the first semiconductor electronic devices. The discovery of crystals' rectifying abilities was made by German physicist Ferdinand Braun in 1874. The first semiconductor diodes, called cat's whisker diodes, developed around 1906, were made of mineral crystals such as galena. Today, most diodes are made of silicon, but other semiconductors such as selenium or germanium are sometimes used.

Main Functions

The most common function of a diode is to allow an electric current to pass in one direction (called the diode's *forward* direction), while blocking current in the opposite direction (the *reverse* direction). Thus, the diode can be viewed as an electronic version of a check valve. This unidirectional behavior is called rectification, and is used to convert alternating current (AC) to direct current (DC), including extraction of modulation from radio signals in radio receivers—these diodes are forms of rectifiers.

However, diodes can have more complicated behavior than this simple on–off action, because of their nonlinear current-voltage characteristics. Semiconductor diodes begin conducting electricity only if a certain threshold voltage or cut-in voltage is present in the forward direction (a state in which the diode is said to be *forward-biased*). The voltage drop across a forward-biased diode varies only a little with the current, and is a function of temperature; this effect can be used as a temperature sensor or as a voltage reference.

A semiconductor diode's current–voltage characteristic can be tailored by selecting the semiconductor materials and the doping impurities introduced into the materials during manufacture. These techniques are used to create special-purpose diodes that perform many different functions. For example, diodes are used to regulate voltage (Zener diodes), to protect circuits from high voltage surges (avalanche diodes), to electronically tune radio and TV receivers (varactor diodes), to generate radio-frequency oscillations (tunnel diodes, Gunn diodes, IMPATT diodes), and to produce light (light-emitting diodes). Tunnel, Gunn and IMPATT diodes exhibit negative resistance, which is useful in microwave and switching circuits.

Diodes, both vacuum and semiconductor, can be used as shot-noise generators.

History

Thermionic (vacuum tube) diodes and solid state (semiconductor) diodes were developed separately, at approximately the same time, in the early 1900s, as radio receiver detectors. Until the 1950s vacuum tube diodes were used more frequently in radios because the early point-contact type semiconductor diodes were less stable. In addition, most receiving sets had vacuum tubes for amplification that could easily have the thermionic diodes included in the tube (for example the 12SQ7 double diode triode), and vacuum tube rectifiers and gas-filled rectifiers were capable of handling some high voltage/high current rectification tasks better than the semiconductor diodes (such as selenium rectifiers) which were available at that time.

Vacuum Tube Diodes

In 1873, Frederick Guthrie discovered the basic principle of operation of thermionic diodes. Guthrie discovered that a positively charged electroscope could be discharged by bringing a grounded piece of white-hot metal close to it (but not actually touching it). The same did not apply to a negatively charged electroscope, indicating that the current flow was only possible in one direction.

Thomas Edison independently rediscovered the principle on February 13, 1880. At the time, Edison was investigating why the filaments of his carbon-filament light bulbs nearly always burned out at the positive-connected end. He had a special bulb made with a metal plate sealed into the glass envelope. Using this device, he confirmed that an invisible current flowed from the glowing filament through the vacuum to the metal plate, but only when the plate was connected to the positive supply.

Edison devised a circuit where his modified light bulb effectively replaced the resistor in a DC voltmeter. Edison was awarded a patent for this invention in 1884. Since there was no apparent practical use for such a device at the time, the patent application was most likely simply a precaution in case someone else did find a use for the so-called Edison effect.

About 20 years later, John Ambrose Fleming (scientific adviser to the Marconi Company and former Edison employee) realized that the Edison effect could be used as a precision radio detector. Fleming patented the first true thermionic diode, the Fleming valve, in Britain on November 16, 1904 (followed by U.S. Patent 803,684 in November 1905).

Solid-State Diodes

In 1874 German scientist Karl Ferdinand Braun discovered the "unilateral conduction" of crystals. Braun patented the crystal rectifier in 1899. Copper oxide and selenium rectifiers were developed for power applications in the 1930s.

Indian scientist Jagadish Chandra Bose was the first to use a crystal for detecting radio waves in 1894. The crystal detector was developed into a practical device for wireless telegraphy by Greenleaf Whittier Pickard, who invented a silicon crystal detector in 1903 and received a patent for it on November 20, 1906. Other experimenters tried a variety of other substances, of which the most widely used was the mineral galena (lead sulfide). Other substances offered slightly better performance, but galena was most widely used because it had the advantage of being cheap and easy to obtain. The crystal detector in these early crystal radio sets consisted of an adjustable wire point-contact, often made of gold or platinum because of their incorrodible nature (the so-called "cat's whisker"), which could be manually moved over the face of the crystal in search of a portion of that mineral with rectifying qualties. This troublesome device was superseded by thermionic diodes (vacuum tubes) by the 1920s, but after high purity semiconductor materials became available, the crystal detector returned to dominant use with the advent, in the 1950s, of inexpensive fixed-germanium diodes. Bell Labs also developed a germanium diode for microwave reception, and AT&T used these in their microwave towers that criss-crossed the nation starting in the late 1940s, carrying telephone and network television signals. Bell Labs did not develop a satisfactory thermionic diode for microwave reception.

Etymology

At the time of their invention, such devices were known as rectifiers. In 1919, the year tetrodes were invented, William Henry Eccles coined the term *diode* from the Greek roots *di* (from δι), meaning 'two', and *ode* eaning 'path'. (However, the word *diode* itself, as well as *triode, tetrode, pentode, hexode*, were already in use as terms of multiplex telegraphy.

Rectifiers

Although all diodes *rectify*, the term 'rectifier' is normally reserved for higher currents and voltages than would normally be found in the rectification of lower power signals; examples include:

* Power supply rectifiers (*half-wave, full-wave, bridge*)
* Flyback diodes

World's Smallest Diode

Researchers from the University of Georgia and Ben-Gurion University of the Negev (BGU) have developed a diode made from a molecule of DNA. Professor Bingqian Xu from the College of Engineering at the University of Georgia and his team took a single DNA molecule made from 11 base pairs and connected it to an electronic circuit a few nanometers in size. When layers of coralyne were inserted between layers of DNA, the current jumped up to 15 times larger negative versus positive, which is necessary for a nano diode.

Thermionic Diodes

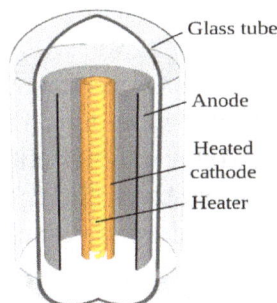

Diode vacuum tube construction

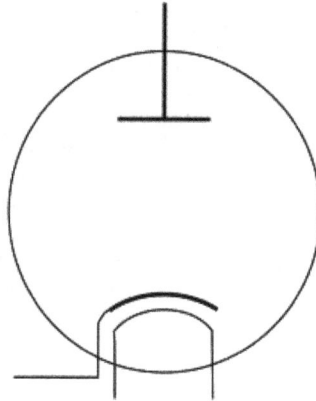

The symbol for an indirect heated vacuum-tube diode. From top to bottom, the components are the anode, the cathode, and the heater filament.

A thermionic diode is a thermionic-valve device (also known as a vacuum tube, tube, or valve), consisting of a sealed evacuated glass envelope containing two electrodes: a cathode heated by a filament, and a plate (anode). Early examples were fairly similar in appearance to incandescent light bulbs.

In operation, a separate current through the filament (heater), a high resistance wire made of nichrome, heats the cathode red hot (800–1000 °C), causing it to release electrons into the vacuum, a process called thermionic emission. The cathode is coated with oxides of alkaline earth metals such as barium and strontium oxides, which have a low work function, to increase the number of electrons emitted. (Some valves use *direct heating*, in which a tungsten filament acts as both heater and cathode.) The alternating voltage to be rectified is applied between the cathode and the concentric plate electrode. When the plate has a positive voltage with respect to the cathode, it electrostatically attracts the electrons from the cathode, so a current of electrons flows through the tube from cathode to plate. However, when the polarity is reversed and the plate has a negative voltage, no current flows, because the cathode electrons are not attracted to it. The unheated plate does not emit any electrons itself. So electrons can only flow through the tube in one direction, from the cathode to the anode plate.

In a mercury-arc valve, an arc forms between a refractory conductive anode and a pool of liquid mercury acting as cathode. Such units were made with ratings up to hundreds of kilowatts, and were important in the development of HVDC power transmission. Some types of smaller thermionic rectifiers had mercury vapor fill to reduce their forward voltage drop and to increase current rating over thermionic hard-vacuum devices.

Throughout the vacuum tube era, valve diodes were used in analog signal applications and as rectifiers in DC power supplies in consumer electronics such as radios, televisions, and sound systems. They were replaced in power supplies beginning in the 1940s by selenium rectifiers and then by semiconductor diodes by the 1960s. Today they are still used in a few high power applications where their ability to withstand transient voltages and their robustness gives them an advantage over semiconductor devices. The recent (2012) resurgence of interest among audiophiles and recording studios in old valve audio gear such as guitar amplifiers and home audio systems has provided a market for the legacy consumer diode valves.

Semiconductor Diodes

Electronic Symbols

The symbol used for a semiconductor diode in a circuit diagram specifies the type of diode. There are alternative symbols for some types of diodes, though the differences are minor. The triangle in the symbols points to the forward direction.

(TVS symbols)	Anode ▷⊦ Cathode	Anode (+) ▷⊦ Cathode (−)
Transient-voltage-suppression diode (TVS)	Diode	Typical diode packages in same alignment as diode symbol. Thin bar depicts the cathode.
Anode ▷⊦ Cathode	Anode ▷⊦ Cathode	Anode ▷⊦ Cathode
Tunnel diode	Light-emitting diode (LED)	Schottky diode
Anode ▷⊦⊦ Cathode	Anode ▷⊦ Cathode	Anode ▷⊦ Cathode
Varicap	Photodiode	Zener diode

A galena cat's-whisker detector, a point-contact diode.

Point-Contact Diodes

A point-contact diode works the same as the junction diodes described below, but its construction is simpler. A pointed metal wire is placed in contact with an n-type semiconductor. Some metal migrates into the semiconductor to make a small p-type region around the contact. The 1N34 germanium version is still used in radio receivers as a detector and occasionally in specialized analog electronics.

Junction Diodes

P–N Junction Diode

A p–n junction diode is made of a crystal of semiconductor, usually silicon, but germanium and gallium arsenide are also used. Impurities are added to it to create a region on one side that contains negative charge carriers (electrons), called an n-type semiconductor, and a region on the other side that contains positive charge carriers (holes), called a p-type semiconductor. When the n-type and p-type materials are attached together, a momentary flow of electrons occur from the n to the p side resulting in a third region between the two where no charge carriers are present. This region is called the depletion region because there are no charge carriers (neither electrons nor holes) in it. The diode's terminals are attached to the n-type and p-regions. The boundary between these two regions, called a p–n junction, is where the action of the diode takes place. When a sufficiently higher electrical potential is applied to the P side (the anode) than to the N side (the cathode), it allows electrons to flow through the depletion region from the N-type side to the P-type side. The junction does not allow the flow of electrons in the opposite direction when the potential is applied in reverse, creating, in a sense, an electrical check valve.

Schottky Diode

Another type of junction diode, the Schottky diode, is formed from a metal–semiconductor junction rather than a p–n junction, which reduces capacitance and increases switching speed.

Current–Voltage Characteristic

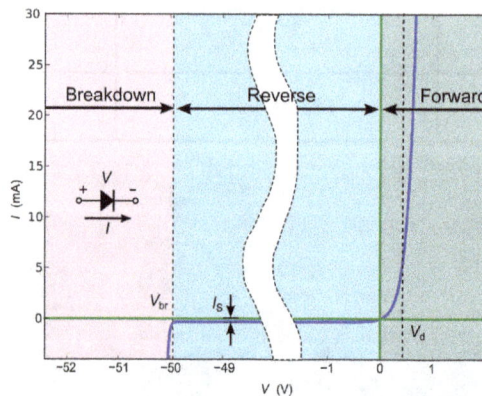

I–V (current vs. voltage) characteristics of a p–n junction diode

A semiconductor diode's behavior in a circuit is given by its current–voltage characteristic, or I–V graph. The shape of the curve is determined by the transport of charge carriers through the so-called *depletion layer* or *depletion region* that exists at the p–n junction between differing semiconductors. When a p–n junction is first created, conduction-band (mobile) electrons from the N-doped region diffuse into the P-doped region where there is a large population of holes (vacant places for electrons) with which the electrons "recombine". When a mobile electron recombines with a hole, both hole and electron vanish, leaving behind an immobile positively charged donor (dopant) on the N side and negatively charged acceptor (dopant) on the P side. The region around the p–n junction becomes depleted of charge carriers and thus behaves as an insulator.

However, the width of the depletion region (called the depletion width) cannot grow without limit. For each electron–hole pair recombination made, a positively charged dopant ion is left behind in the N-doped region, and a negatively charged dopant ion is created in the P-doped region. As recombination proceeds and more ions are created, an increasing electric field develops through the depletion zone that acts to slow and then finally stop recombination. At this point, there is a "built-in" potential across the depletion zone.

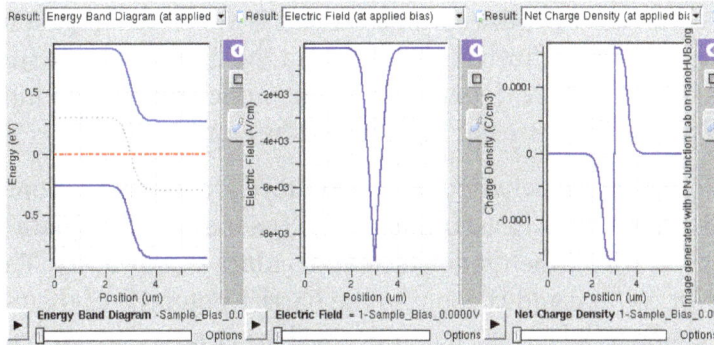

A PN junction diode in forward bias mode, the depletion width decreases. Both p and n junctions are doped at a 1e15/cm3 doping level, leading to built-in potential of ~0.59V. Observe the different Quasi Fermi levels for conduction band and valence band in n and p regions (red curves).

Reverse Bias

If an external voltage is placed across the diode with the same polarity as the built-in potential, the depletion zone continues to act as an insulator, preventing any significant electric current flow (unless electron–hole pairs are actively being created in the junction by, for instance, light; photodiode). This is called the *reverse bias* phenomenon.

Forward Bias

However, if the polarity of the external voltage opposes the built-in potential, recombination can once again proceed, resulting in a substantial electric current through the p–n junction (i.e. substantial numbers of electrons and holes recombine at the junction). For silicon diodes, the built-in potential is approximately 0.7 V (0.3 V for germanium and 0.2 V for Schottky). Thus, if an external voltage greater than and opposite to the built-in voltage is applied, a current will flow and the diode is said to be "turned on" as it has been given an external *forward bias*. The diode is commonly said to have a forward "threshold" voltage, above which it conducts and below which conduction stops. However, this is only an approximation as the forward characteristic is according to the Shockley equation absolutely smooth.

A diode's I–V characteristic can be approximated by four regions of operation:

1. At very large reverse bias, beyond the peak inverse voltage or PIV, a process called reverse breakdown occurs that causes a large increase in current (i.e., a large number of electrons and holes are created at, and move away from the p–n junction) that usually damages the device permanently. The avalanche diode is deliberately designed for use in that manner. In the Zener diode, the concept of PIV is not applicable. A Zener diode contains a heavily doped p–n junction allowing electrons to tunnel from the valence band of the p-type

material to the conduction band of the n-type material, such that the reverse voltage is "clamped" to a known value (called the *Zener voltage*), and avalanche does not occur. Both devices, however, do have a limit to the maximum current and power they can withstand in the clamped reverse-voltage region. Also, following the end of forward conduction in any diode, there is reverse current for a short time. The device does not attain its full blocking capability until the reverse current ceases.

2. For a bias less than the PIV, the reverse current is very small. For a normal P–N rectifier diode, the reverse current through the device in the micro-ampere (μA) range is very low. However, this is temperature dependent, and at sufficiently high temperatures, a substantial amount of reverse current can be observed (mA or more).

3. With a small forward bias, where only a small forward current is conducted, the current–voltage curve is exponential in accordance with the ideal diode equation. There is a definite forward voltage at which the diode starts to conduct significantly. This is called the *knee voltage* or *cut-in voltage* and is equal to the barrier potential of the p-n junction. This is a feature of the exponential curve, and appears sharper on a current scale more compressed than in the diagram shown here.

At larger forward currents the current-voltage curve starts to be dominated by the ohmic resistance of the bulk semiconductor. The curve is no longer exponential, it is asymptotic to a straight line whose slope is the bulk resistance. This region is particularly important for power diodes. The diode can be modeled as an ideal diode in series with a fixed resistor.In a small silicon diode operating at its rated currents, the voltage drop is about 0.6 to 0.7 volts. The value is different for other diode types—Schottky diodes can be rated as low as 0.2 V, germanium diodes 0.25 to 0.3 V, and red or blue light-emitting diodes (LEDs) can have values of 1.4 V and 4.0 V respectively.

At higher currents the forward voltage drop of the diode increases. A drop of 1 V to 1.5 V is typical at full rated current for power diodes.

Shockley Diode Equation

The *Shockley ideal diode equation* or the *diode law* (named after transistor co-inventor William Bradford Shockley) gives the I–V characteristic of an ideal diode in either forward or reverse bias (or no bias). The following equation is called the *Shockley ideal diode equation* when n, the ideality factor, is set equal to 1 :

$$I = I_S \left(e^{\frac{V_D}{nV_T}} - 1 \right)$$

where

I is the diode current,

I_S is the reverse bias saturation current (or scale current),

V_D is the voltage across the diode,

V_T is the thermal voltage, and

n is the *ideality factor*, also known as the *quality factor* or sometimes *emission coefficient*. The ideality factor n typically varies from 1 to 2 (though can in some cases be higher), depending on the fabrication process and semiconductor material and is set equal to 1 for the case of an "ideal" diode (thus the n is sometimes omitted). The ideality factor was added to account for imperfect junctions as observed in real transistors. The factor mainly accounts for carrier recombination as the charge carriers cross the depletion region.

The thermal voltage V_T is approximately 25.85 mV at 300 K, a temperature close to "room temperature" commonly used in device simulation software. At any temperature it is a known constant defined by:

$$V_T = \frac{kT}{q},$$

where k is the Boltzmann constant, T is the absolute temperature of the p–n junction, and q is the magnitude of charge of an electron (the elementary charge).

The reverse saturation current, I_S, is not constant for a given device, but varies with temperature; usually more significantly than V_T, so that V_D typically decreases as T increases.

The *Shockley ideal diode equation* or the *diode law* is derived with the assumption that the only processes giving rise to the current in the diode are drift (due to electrical field), diffusion, and thermal recombination–generation (R–G) (this equation is derived by setting n = 1 above). It also assumes that the R–G current in the depletion region is insignificant. This means that the *Shockley ideal diode equation* doesn't account for the processes involved in reverse breakdown and photon-assisted R–G. Additionally, it doesn't describe the "leveling off" of the I–V curve at high forward bias due to internal resistance. Introducing the ideality factor, n, accounts for recombination and generation of carriers.

Under *reverse bias* voltages the exponential in the diode equation is negligible, and the current is a constant (negative) reverse current value of $-I_S$. The reverse *breakdown region* is not modeled by the Shockley diode equation.

For even rather small *forward bias* voltages the exponential is very large, since the thermal voltage is very small in comparison. The subtracted '1' in the diode equation is then negligible and the forward diode current can be approximated by

$$I = I_S e^{\frac{V_D}{nV_T}}$$

The use of the diode equation in circuit problems is illustrated in the article on diode modeling.

Small-Signal Behavior

For circuit design, a small-signal model of the diode behavior often proves useful. A specific example of diode modeling is discussed in the article on small-signal circuits.

Reverse-Recovery Effect

Following the end of forward conduction in a p–n type diode, a reverse current can flow for a short time. The device does not attain its blocking capability until the mobile charge in the junction is depleted.

The effect can be significant when switching large currents very quickly. A certain amount of "reverse recovery time" t_r (on the order of tens of nanoseconds to a few microseconds) may be required to remove the reverse recovery charge Q_r from the diode. During this recovery time, the diode can actually conduct in the reverse direction. This might give rise to a large constant current in the reverse direction for a short time while the diode is reverse biased. The magnitude of such a reverse current is determined by the operating circuit (i.e., the series resistance) and the diode is said to be in the storage-phase. In certain real-world cases it is important to consider the losses that are incurred by this non-ideal diode effect. However, when the slew rate of the current is not so severe (e.g. Line frequency) the effect can be safely ignored. For most applications, the effect is also negligible for Schottky diodes.

The reverse current ceases abruptly when the stored charge is depleted; this abrupt stop is exploited in step recovery diodes for generation of extremely short pulses.

Types of Semiconductor Diode

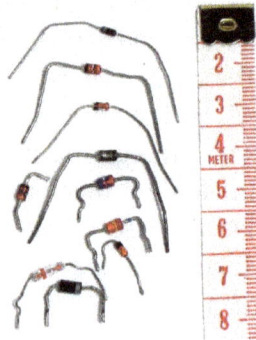

Several types of diodes. The scale is centimeters.

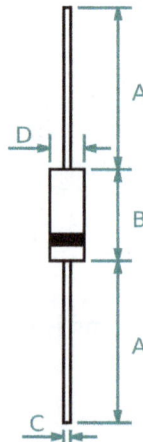

Typical datasheet drawing showing the dimensions of a DO-41 diode package

There are several types of p–n junction diodes, which emphasize either a different physical aspect of a diode often by geometric scaling, doping level, choosing the right electrodes, are just an application of a diode in a special circuit, or are really different devices like the Gunn and laser diode and the MOSFET:

Normal (p–n) diodes, which operate as described above, are usually made of doped silicon or, more rarely, germanium. Before the development of silicon power rectifier diodes, cuprous oxide and later selenium was used. Their low efficiency required a much higher forward voltage to be applied (typically 1.4 to 1.7 V per "cell", with multiple cells stacked so as to increase the peak inverse voltage rating for application in high voltage rectifiers), and required a large heat sink (often an extension of the diode's metal substrate), much larger than the later silicon diode of the same current ratings would require. The vast majority of all diodes are the p–n diodes found in CMOS integrated circuits, which include two diodes per pin and many other internal diodes.

Avalanche Diodes

These are diodes that conduct in the reverse direction when the reverse bias voltage exceeds the breakdown voltage. These are electrically very similar to Zener diodes (and are often mistakenly called Zener diodes), but break down by a different mechanism: the *avalanche effect*. This occurs when the reverse electric field applied across the p–n junction causes a wave of ionization, reminiscent of an avalanche, leading to a large current. Avalanche diodes are designed to break down at a well-defined reverse voltage without being destroyed. The difference between the avalanche diode (which has a reverse breakdown above about 6.2 V) and the Zener is that the channel length of the former exceeds the mean free path of the electrons, resulting in many collisions between them on the way through the channel. The only practical difference between the two types is they have temperature coefficients of opposite polarities.

Cat's whisker or crystal diodes

These are a type of point-contact diode. The cat's whisker diode consists of a thin or sharpened metal wire pressed against a semiconducting crystal, typically galena or a piece of coal. The wire forms the anode and the crystal forms the cathode. Cat's whisker diodes were also called crystal diodes and found application in the earliest radios called crystal radio receivers. Cat's whisker diodes are generally obsolete, but may be available from a few manufacturers.

Constant current diodes

These are actually JFETs with the gate shorted to the source, and function like a two-terminal current-limiting analog to the voltage-limiting Zener diode. They allow a current through them to rise to a certain value, and then level off at a specific value. Also called *CLDs*, *constant-current diodes*, *diode-connected transistors*, or *current-regulating diodes*.

Esaki or tunnel diodes

These have a region of operation showing negative resistance caused by quantum tunneling, allowing amplification of signals and very simple bistable circuits. Because of the high carrier concentration, tunnel diodes are very fast, may be used at low (mK) temperatures, high magnetic fields, and in high radiation environments. Because of these properties, they are often used in spacecraft.

Gunn diodes

These are similar to tunnel diodes in that they are made of materials such as GaAs or InP that exhibit a region of negative differential resistance. With appropriate biasing, dipole domains form and travel across the diode, allowing high frequency microwave oscillators to be built.

Light-emitting diodes (LEDs)

In a diode formed from a direct band-gap semiconductor, such as gallium arsenide, charge carriers that cross the junction emit photons when they recombine with the majority carrier on the other side. Depending on the material, wavelengths (or colors) from the infrared to the near ultraviolet may be produced. The forward potential of these diodes depends on the wavelength of the emitted photons: 2.1 V corresponds to red, 4.0 V to violet. The first LEDs were red and yellow, and higher-frequency diodes have been developed over time. All LEDs produce incoherent, narrow-spectrum light; "white" LEDs are actually combinations of three LEDs of a different color, or a blue LED with a yellow scintillator coating. LEDs can also be used as low-efficiency photodiodes in signal applications. An LED may be paired with a photodiode or phototransistor in the same package, to form an opto-isolator.

Laser diodes

When an LED-like structure is contained in a resonant cavity formed by polishing the parallel end faces, a laser can be formed. Laser diodes are commonly used in optical storage devices and for high speed optical communication.

Thermal diodes

This term is used both for conventional p–n diodes used to monitor temperature because of their varying forward voltage with temperature, and for Peltier heat pumps for thermoelectric heating and cooling. Peltier heat pumps may be made from semiconductor, though they do not have any rectifying junctions, they use the differing behaviour of charge carriers in N and P type semiconductor to move heat.

Perun's diodes

This is a special type of voltage-surge protection diode. It is characterized by the symmetrical voltage-current characteristic, similar to DIAC. It has much faster response time however, that's why it is used in demanding applications.

Photodiodes

All semiconductors are subject to optical charge carrier generation. This is typically an undesired effect, so most semiconductors are packaged in light blocking material. Photodiodes are intended to sense light(photodetector), so they are packaged in materials that allow light to pass, and are usually PIN (the kind of diode most sensitive to light). A photodiode can be used in solar cells, in photometry, or in optical communications. Multiple photodiodes may be packaged in a single device, either as a linear array or as a two-dimensional array. These arrays should not be confused with charge-coupled devices.

PIN diodes

A PIN diode has a central un-doped, or *intrinsic*, layer, forming a p-type/intrinsic/n-type structure. They are used as radio frequency switches and attenuators. They are also used as large-volume, ionizing-radiation detectors and as photodetectors. PIN diodes are also used in power electronics, as their central layer can withstand high voltages. Furthermore, the PIN structure can be found in many power semiconductor devices, such as IGBTs, power MOSFETs, and thyristors.

Schottky diodes

Schottky diodes are constructed from a metal to semiconductor contact. They have a lower forward voltage drop than p–n junction diodes. Their forward voltage drop at forward currents of about 1 mA is in the range 0.15 V to 0.45 V, which makes them useful in voltage clamping applications and prevention of transistor saturation. They can also be used as low loss rectifiers, although their reverse leakage current is in general higher than that of other diodes. Schottky diodes are majority carrier devices and so do not suffer from minority carrier storage problems that slow down many other diodes—so they have a faster reverse recovery than p–n junction diodes. They also tend to have much lower junction capacitance than p–n diodes, which provides for high switching speeds and their use in high-speed circuitry and RF devices such as switched-mode power supply, mixers, and detectors.

Super barrier diodes

Super barrier diodes are rectifier diodes that incorporate the low forward voltage drop of the Schottky diode with the surge-handling capability and low reverse leakage current of a normal p–n junction diode.

Gold-doped diodes

As a dopant, gold (or platinum) acts as recombination centers, which helps a fast recombination of minority carriers. This allows the diode to operate at signal frequencies, at the expense of a higher forward voltage drop. Gold-doped diodes are faster than other p–n diodes (but not as fast as Schottky diodes). They also have less reverse-current leakage than Schottky diodes (but not as good as other p–n diodes). A typical example is the 1N914.

Snap-off or Step recovery diodes

The term *step recovery* relates to the form of the reverse recovery characteristic of these devices. After a forward current has been passing in an SRD and the current is interrupted or reversed, the reverse conduction will cease very abruptly (as in a step waveform). SRDs can, therefore, provide very fast voltage transitions by the very sudden disappearance of the charge carriers.

Stabistors or *Forward Reference Diodes*

> The term *stabistor* refers to a special type of diodes featuring extremely stable forward voltage characteristics. These devices are specially designed for low-voltage stabilization applications requiring a guaranteed voltage over a wide current range and highly stable over temperature.

Transient voltage suppression diode (TVS)

> These are avalanche diodes designed specifically to protect other semiconductor devices from high-voltage transients. Their p–n junctions have a much larger cross-sectional area than those of a normal diode, allowing them to conduct large currents to ground without sustaining damage.

Varicap or varactor diodes

> These are used as voltage-controlled capacitors. These are important in PLL (phase-locked loop) and FLL (frequency-locked loop) circuits, allowing tuning circuits, such as those in television receivers, to lock quickly on to the frequency. They also enabled tunable oscillators in early discrete tuning of radios, where a cheap and stable, but fixed-frequency, crystal oscillator provided the reference frequency for a voltage-controlled oscillator.

Zener diodes

> These can be made to conduct in reverse bias (backward), and are correctly termed reverse breakdown diodes. This effect, called Zener breakdown, occurs at a precisely defined voltage, allowing the diode to be used as a precision voltage reference. The term Zener diode is colloquially applied to several types of breakdown diodes, but strictly speaking Zener diodes have a breakdown voltage of below 5 volts, whilst avalanche diodes are used for breakdown voltages above that value. In practical voltage reference circuits, Zener and switching diodes are connected in series and opposite directions to balance the temperature coefficient response of the diodes to near-zero. Some devices labeled as high-voltage Zener diodes are actually avalanche diodes. Two (equivalent) Zeners in series and in reverse order, in the same package, constitute a transient absorber (or Transorb, a registered trademark).

Other uses for semiconductor diodes include the sensing of temperature, and computing analog logarithms.

Numbering and Coding Schemes

There are a number of common, standard and manufacturer-driven numbering and coding schemes for diodes; the two most common being the EIA/JEDEC standard and the European Pro Electron standard:

EIA/JEDEC

The standardized 1N-series numbering *EIA370* system was introduced in the US by EIA/JEDEC

(Joint Electron Device Engineering Council) about 1960. Most diodes have a 1-prefix designation (e.g., 1N4003). Among the most popular in this series were: 1N34A/1N270 (germanium signal), 1N914/1N4148 (silicon signal), 1N4001-1N4007 (silicon 1A power rectifier) and 1N54xx (silicon 3A power rectifier)

JIS

The JIS semiconductor designation system has all semiconductor diode designations starting with "1S".

Pro Electron

The European Pro Electron coding system for active components was introduced in 1966 and comprises two letters followed by the part code. The first letter represents the semiconductor material used for the component (A = germanium and B = silicon) and the second letter represents the general function of the part (for diodes, A = low-power/signal, B = variable capacitance, X = multiplier, Y = rectifier and Z = voltage reference); for example:

- AA-series germanium low-power/signal diodes (e.g., AA119)

- BA-series silicon low-power/signal diodes (e.g., BAT18 silicon RF switching diode)

- BY-series silicon rectifier diodes (e.g., BY127 1250V, 1A rectifier diode)

- BZ-series silicon Zener diodes (e.g., BZY88C4V7 4.7V Zener diode)

Other common numbering / coding systems (generally manufacturer-driven) include:

- GD-series germanium diodes (e.g., GD9) – this is a very old coding system

- OA-series germanium diodes (e.g., OA47) – a coding sequence developed by Mullard, a UK company

As well as these common codes, many manufacturers or organisations have their own systems too – for example:

- HP diode 1901-0044 = JEDEC 1N4148

- UK military diode CV448 = Mullard type OA81 = GEC type GEX23

Related Devices

- Rectifier

- Transistor

- Thyristor or silicon controlled rectifier (SCR)

- TRIAC

- DIAC

- Varistor

In optics, an equivalent device for the diode but with laser light would be the Optical isolator, also known as an Optical Diode, that allows light to only pass in one direction. It uses a Faraday rotator as the main component.

Applications

Radio Demodulation

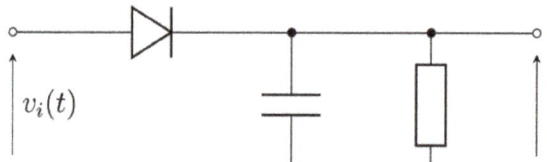

A simple envelope demodulator circuit.

The first use for the diode was the demodulation of amplitude modulated (AM) radio broadcasts. The history of this discovery is treated in depth in the radio article. In summary, an AM signal consists of alternating positive and negative peaks of a radio carrier wave, whose amplitude or envelope is proportional to the original audio signal. The diode (originally a crystal diode) rectifies the AM radio frequency signal, leaving only the positive peaks of the carrier wave. The audio is then extracted from the rectified carrier wave using a simple filter and fed into an audio amplifier or transducer, which generates sound waves.

Power Conversion

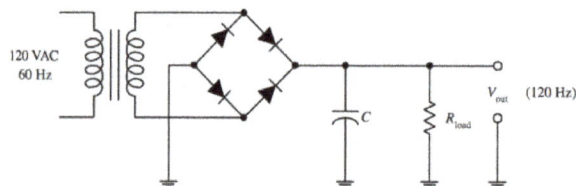

Schematic of basic AC-to-DC power supply

Rectifiers are constructed from diodes, where they are used to convert alternating current (AC) electricity into direct current (DC). Automotive alternators are a common example, where the diode, which rectifies the AC into DC, provides better performance than the commutator or earlier, dynamo. Similarly, diodes are also used in *Cockcroft–Walton voltage multipliers* to convert AC into higher DC voltages.

Over-Voltage Protection

Diodes are frequently used to conduct damaging high voltages away from sensitive electronic devices. They are usually reverse-biased (non-conducting) under normal circumstances. When the voltage rises above the normal range, the diodes become forward-biased (conducting). For example, diodes are used in (stepper motor and H-bridge) motor controller and relay circuits to de-energize coils rapidly without the damaging voltage spikes that would otherwise occur. (A diode used in such an application is called a flyback diode). Many integrated circuits also incorporate diodes on the connection pins to prevent external voltages from damaging their sensitive transistors. Specialized diodes are used to protect from over-voltages at higher power.

Logic Gates

Diodes can be combined with other components to construct AND and OR logic gates. This is referred to as diode logic.

Ionizing Radiation Detectors

In addition to light, mentioned above, semiconductor diodes are sensitive to more energetic radiation. In electronics, cosmic rays and other sources of ionizing radiation cause noise pulses and single and multiple bit errors. This effect is sometimes exploited by particle detectors to detect radiation. A single particle of radiation, with thousands or millions of electron volts of energy, generates many charge carrier pairs, as its energy is deposited in the semiconductor material. If the depletion layer is large enough to catch the whole shower or to stop a heavy particle, a fairly accurate measurement of the particle's energy can be made, simply by measuring the charge conducted and without the complexity of a magnetic spectrometer, etc. These semiconductor radiation detectors need efficient and uniform charge collection and low leakage current. They are often cooled by liquid nitrogen. For longer-range (about a centimetre) particles, they need a very large depletion depth and large area. For short-range particles, they need any contact or un-depleted semiconductor on at least one surface to be very thin. The back-bias voltages are near breakdown (around a thousand volts per centimetre). Germanium and silicon are common materials. Some of these detectors sense position as well as energy. They have a finite life, especially when detecting heavy particles, because of radiation damage. Silicon and germanium are quite different in their ability to convert gamma rays to electron showers.

Semiconductor detectors for high-energy particles are used in large numbers. Because of energy loss fluctuations, accurate measurement of the energy deposited is of less use.

Temperature Measurements

A diode can be used as a temperature measuring device, since the forward voltage drop across the diode depends on temperature, as in a silicon bandgap temperature sensor. From the Shockley ideal diode equation given above, it might *appear* that the voltage has a *positive* temperature coefficient (at a constant current), but usually the variation of the reverse saturation current term is more significant than the variation in the thermal voltage term. Most diodes therefore have a *negative* temperature coefficient, typically −2 mV/°C for silicon diodes. The temperature coefficient is approximately constant for temperatures above about 20 kelvins. Some graphs are given for 1N400x series, and CY7 cryogenic temperature sensor.

Current Steering

Diodes will prevent currents in unintended directions. To supply power to an electrical circuit during a power failure, the circuit can draw current from a battery. An uninterruptible power supply may use diodes in this way to ensure that current is only drawn from the battery when necessary. Likewise, small boats typically have two circuits each with their own battery/batteries: one used for engine starting; one used for domestics. Normally, both are charged from a single alternator, and a heavy-duty split-charge diode is used to prevent the higher-charge battery (typically the engine battery) from discharging through the lower-charge battery when the alternator is not running.

Diodes are also used in electronic musical keyboards. To reduce the amount of wiring needed in electronic musical keyboards, these instruments often use keyboard matrix circuits. The keyboard controller scans the rows and columns to determine which note the player has pressed. The problem with matrix circuits is that, when several notes are pressed at once, the current can flow backwards through the circuit and trigger "phantom keys" that cause "ghost" notes to play. To avoid triggering unwanted notes, most keyboard matrix circuits have diodes soldered with the switch under each key of the musical keyboard. The same principle is also used for the switch matrix in solid-state pinball machines.

Waveform Clipper

Diodes can be used to limit the positive or negative excursion of a signal to a prescribed voltage.

Clamper

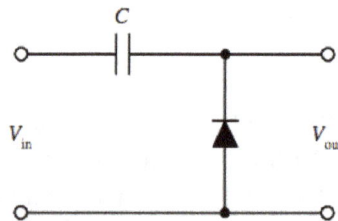

This simple diode clamp will clamp the negative peaks of the incoming waveform to the common rail voltage

A diode clamp circuit can take a periodic alternating current signal that oscillates between positive and negative values, and vertically displace it such that either the positive, or the negative peaks occur at a prescribed level. The clamper does not restrict the peak-to-peak excursion of the signal, it moves the whole signal up or down so as to place the peaks at the reference level.

Abbreviations

Diodes are usually referred to as D for diode on PCBs. Sometimes the abbreviation CR for *crystal rectifier* is used.

Zener Diode

A Zener diode allows current to flow from its anode to its cathode like a normal semiconductor diode, but it also permits current to flow in the reverse direction when its "Zener voltage" is reached. Zener diodes have a highly doped p-n junction. Normal diodes will also break down with a reverse voltage but the voltage and sharpness of the knee are not as well defined as for a Zener diode. Also normal diodes are not designed to operate in the breakdown region, but Zener diodes can reliably operate in this region.

The device was named after Clarence Melvin Zener, who discovered the Zener effect. Zener reverse breakdown is due to electron quantum tunnelling caused by a high strength electric field.

However, many diodes described as "Zener" diodes rely instead on avalanche breakdown. Both breakdown types are used in Zener diodes with the Zener effect predominating under 5.6 V and avalanche breakdown above.

Zener diodes are widely used in electronic equipment of all kinds and are one of the basic building blocks of electronic circuits. They are used to generate low power stabilized supply rails from a higher voltage and to provide reference voltages for circuits, especially stabilized power supplies. They are also used to protect circuits from over-voltage, especially electrostatic discharge (ESD).

Operation

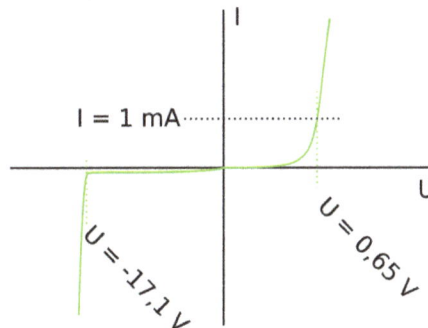

Current-voltage characteristic of a Zener diode with a breakdown voltage of 17 volts. Notice the change of voltage scale between the forward biased (positive) direction and the reverse biased (negative) direction.

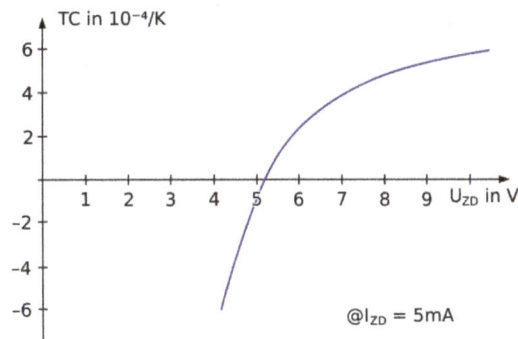

Temperature coefficient of Zener voltage against nominal Zener voltage.

A conventional solid-state diode allows significant current if it is reverse-biased above its reverse breakdown voltage. When the reverse bias breakdown voltage is exceeded, a conventional diode is subject to high current due to avalanche breakdown. Unless this current is limited by circuitry, the diode may be permanently damaged due to overheating. A Zener diode exhibits almost the same properties, except the device is specially designed so as to have a reduced breakdown voltage, the so-called Zener voltage. By contrast with the conventional device, a reverse-biased Zener diode exhibits a controlled breakdown and allows the current to keep the voltage across the Zener diode close to the Zener breakdown voltage. For example, a diode with a Zener breakdown voltage of 3.2 V exhibits a voltage drop of very nearly 3.2 V across a wide range of reverse currents. The Zener diode is therefore ideal for applications such as the generation of a reference voltage (e.g. for an amplifier stage), or as a voltage stabilizer for low-current applications.

Another mechanism that produces a similar effect is the avalanche effect as in the avalanche diode. The two types of diode are in fact constructed the same way and both effects are present in diodes of this type. In silicon diodes up to about 5.6 volts, the Zener effect is the predominant effect and shows a marked negative temperature coefficient. Above 5.6 volts, the avalanche effect becomes predominant and exhibits a positive temperature coefficient.

In a 5.6 V diode, the two effects occur together, and their temperature coefficients nearly cancel each other out, thus the 5.6 V diode is useful in temperature-critical applications. An alternative, which is used for voltage references that need to be highly stable over long periods of time, is to use a Zener diode with a temperature coefficient (TC) of +2 mV/°C (breakdown voltage 6.2–6.3 V) connected in series with a forward-biased silicon diode (or a transistor B-E junction) manufactured on the same chip. The forward-biased diode has a temperature coefficient of –2 mV/°C, causing the TCs to cancel out.

Modern manufacturing techniques have produced devices with voltages lower than 5.6 V with negligible temperature coefficients, but as higher-voltage devices are encountered, the temperature coefficient rises dramatically. A 75 V diode has 10 times the coefficient of a 12 V diode.

Zener and avalanche diodes, regardless of breakdown voltage, are usually marketed under the umbrella term of "Zener diode".

Under 5.6 V, where the Zener effect dominates, the IV curve near breakdown is much more rounded, which calls for more care in targeting its biasing conditions. The IV curve for Zeners above 5.6 V (being dominated by Avalanche), is much sharper at breakdown.

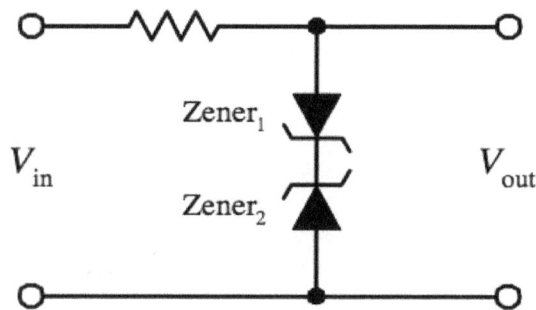

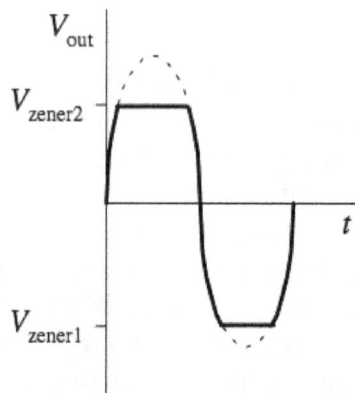

Examples of a Waveform Clipper

Waveform Clipper

Two Zener diodes facing each other in series will act to clip both halves of an input signal. Waveform clippers can be used to not only reshape a signal, but also to prevent voltage spikes from affecting circuits that are connected to the power supply.

Voltage Shifter

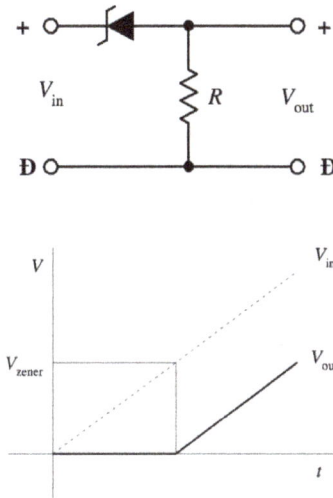

Examples of a Voltage Shifter

A Zener diode can be applied to a circuit with a resistor to act as a voltage shifter. This circuit lowers the output voltage by a quantity that is equal to the Zener diode's breakdown voltage.

Voltage Regulator

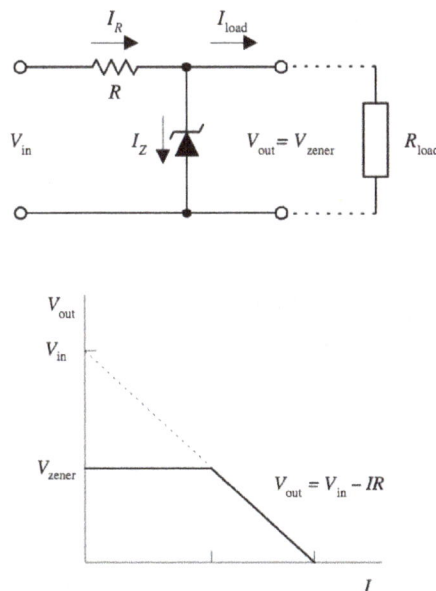

Examples of a Voltage Regulator

A Zener diode can be applied in a voltage regulator circuit to regulate the voltage applied to a load, such as in a linear regulator.

Construction

The Zener diode's operation depends on the heavy doping of its p-n junction. The depletion region formed in the diode is very thin (<1 µm) and the electric field is consequently very high (about 500 kV/m) even for a small reverse bias voltage of about 5 V, allowing electrons to tunnel from the valence band of the p-type material to the conduction band of the n-type material.

At the atomic scale, this tunneling corresponds to the transport of valence band electrons into the empty conduction band states; as a result of the reduced barrier between these bands and high electric fields that are induced due to the relatively high levels of doping on both sides. The breakdown voltage can be controlled quite accurately in the doping process. While tolerances within 0.07% are available, the most widely used tolerances are 5% and 10%. Breakdown voltage for commonly available Zener diodes can vary widely from 1.2 volts to 200 volts.

For diodes that are lightly doped the breakdown is dominated by the avalanche effect rather than the Zener effect. Consequently, the breakdown voltage is higher (over 5.6 V) for these devices.

Surface Zeners

The emitter-base junction of a bipolar NPN transistor behaves as a Zener diode, with breakdown voltage at about 6.8 V for common bipolar processes and about 10 V for lightly doped base regions in BiCMOS processes. Older processes with poor control of doping characteristics had the variation of Zener voltage up to ±1 V, newer processes using ion implantation can achieve no more than ±0.25 V. The NPN transistor structure can be employed as a *surface Zener diode*, with collector and emitter connected together as its cathode and base region as anode. In this approach the base doping profile usually narrows towards the surface, creating a region with intensified electric field where the avalanche breakdown occurs. The hot carriers produced by acceleration in the intense field sometime shoot into the oxide layer above the junction and become trapped there. The accumulation of trapped charges can then cause 'Zener walkout', a corresponding change of the Zener voltage of the junction. The same effect can be achieved by radiation damage.

The emitter-base Zener diodes can handle only smaller currents as the energy is dissipated in the base depletion region which is very small. Higher amount of dissipated energy (higher current for longer time, or a short very high current spike) causes thermal damage to the junction and/or its contacts. Partial damage of the junction can shift its Zener voltage. Total destruction of the Zener junction by overheating it and causing migration of metallization across the junction ("spiking") can be used intentionally as a 'Zener zap' antifuse.

Subsurface Zeners

A subsurface Zener diode, also called 'buried Zener', is a device similar to the Surface Zener, but with the avalanche region located deeper in the structure, typically several micrometers below the oxide. The hot carriers then lose energy by collisions with the semiconductor lattice before reaching the oxide layer and cannot be trapped there. The Zener walkout phenomenon therefore does not occur here, and the buried Zeners have voltage constant over their entire lifetime. Most buried Zeners have breakdown voltage of 5–7 volts. Several different junction structures are used.

Uses

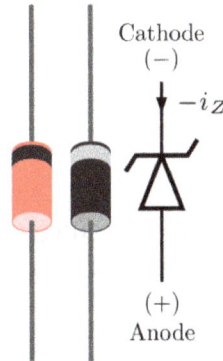

Zener diode shown with typical packages. *Reverse* current $-i_Z$ is shown.

Zener diodes are widely used as voltage references and as shunt regulators to regulate the voltage across small circuits. When connected in parallel with a variable voltage source so that it is reverse biased, a Zener diode conducts when the voltage reaches the diode's reverse breakdown voltage. From that point on, the relatively low impedance of the diode keeps the voltage across the diode at that value.

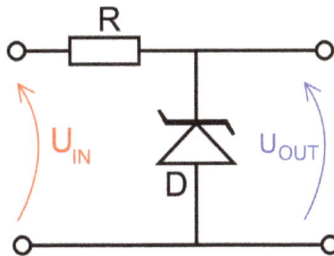

In this circuit, a typical voltage reference or regulator, an input voltage, U_{IN}, is regulated down to a stable output voltage U_{OUT}. The breakdown voltage of diode D is stable over a wide current range and holds U_{OUT} relatively constant even though the input voltage may fluctuate over a fairly wide range. Because of the low impedance of the diode when operated like this, resistor R is used to limit current through the circuit.

In the case of this simple reference, the current flowing in the diode is determined using Ohm's law and the known voltage drop across the resistor R;

$$I_{diode} = \frac{U_{IN} - U_{OUT}}{R_\Omega}$$

The value of R must satisfy two conditions :

- R must be small enough that the current through D keeps D in reverse breakdown. The value of this current is given in the data sheet for D. For example, the common BZX79C5V6 device, a 5.6 V 0.5 W Zener diode, has a recommended reverse current of 5 mA. If insufficient current exists through D, then U_{OUT} is unregulated and less than the nominal break-down voltage (this differs to voltage-regulator tubes where the output voltage will be higher than nominal and could rise as high as U_{IN}). When calculating R, allowance must be made for any current through the external load, not shown in this diagram, connected across U_{OUT}.

- R must be large enough that the current through D does not destroy the device. If the current through D is I_D, its breakdown voltage V_B and its maximum power dissipation P_{MAX} correlate as such: $I_D V_B < P_{MAX}$..

A load may be placed across the diode in this reference circuit, and as long as the Zener stays in reverse breakdown, the diode provides a stable voltage source to the load. Zener diodes in this configuration are often used as stable references for more advanced voltage regulator circuits.

Shunt regulators are simple, but the requirements that the ballast resistor be small enough to avoid excessive voltage drop during worst-case operation (low input voltage concurrent with high load current) tends to leave a lot of current flowing in the diode much of the time, making for a fairly wasteful regulator with high quiescent power dissipation, only suitable for smaller loads.

These devices are also encountered, typically in series with a base-emitter junction, in transistor stages where selective choice of a device centered around the avalanche or Zener point can be used to introduce compensating temperature co-efficient balancing of the transistor p–n junction. An example of this kind of use would be a DC error amplifier used in a regulated power supply circuit feedback loop system.

Zener diodes are also used in surge protectors to limit transient voltage spikes.

Another application of the Zener diode is the use of noise caused by its avalanche breakdown in a random number generator.

Tunnel Diode

10mA germanium tunnel diode mounted in test fixture of Tektronix 571 curve tracer

A tunnel diode or Esaki diode is a type of semiconductor that is capable of very fast operation, well into the microwave frequency region, made possible by the use of the quantum mechanical effect called tunneling.

It was invented in August 1957 by Leo Esaki when he was with Tokyo Tsushin Kogyo, now known as Sony. In 1973 he received the Nobel Prize in Physics, jointly with Brian Josephson, for discovering the electron tunneling effect used in these diodes. Robert Noyce independently came up with the idea of a tunnel diode while working for William Shockley, but was discouraged from pursuing it.

These diodes have a heavily doped p–n junction that is about 10 nm (100 Å) wide. The heavy doping results in a broken band gap, where conduction band electron states on the n-side are more or less aligned with valence band hole states on the p-side.

Tunnel diodes were first manufactured by Sony in 1957 followed by General Electric and other companies from about 1960, and are still made in low volume today. Tunnel diodes are usually made from germanium, but can also be made from gallium arsenide and silicon materials. They are used in frequency converters and detectors. They have negative differential resistance in part of their operating range, and therefore are also used as oscillators, amplifiers, and in switching circuits using hysteresis.

Figure 6: 8–12 GHz tunnel diode amplifier, circa 1970

In 1977, the Intelsat V satellite receiver used a microstrip tunnel diode amplifier (TDA) front-end in the 14 to 15.5 GHz frequency band. Such amplifiers were considered state-of-the-art, with better performance at high frequencies than any transistor-based front end.

The highest frequency room-temperature solid-state oscillators are based on the resonant-tunneling diode (RTD).

There is another type of tunnel diode called a metal–insulator–metal (MIM) diode, but its present application appears to be limited to research environments due to inherent sensitivities. There is also a metal–insulator–insulator–metal MIIM diode which has an additional insulator layer. The additional layer allows *step tunneling* for precise diode control.

Forward Bias Operation

Under normal forward bias operation, as voltage begins to increase, electrons at first tunnel through the very narrow p–n junction barrier and fill electron states in the conduction band on the n-side which become aligned with empty valence band hole states on the p-side of the p–n junction. As voltage increases further, these states become increasingly misaligned and the current drops. This is called *negative resistance* because current decreases with increasing voltage. As voltage increases yet further, the diode begins to operate as a normal diode, where electrons travel by conduction across the p–n junction, and no longer by tunneling through the p–n junction barrier. The most important operating region for a tunnel diode is the negative resistance region. Its graph is different from normal p-n junction diode.

Reverse Bias Operation

When used in the reverse direction, tunnel diodes are called back diodes (or backward diodes) and can act as fast rectifiers with zero offset voltage and extreme linearity for power signals (they have an accurate square law characteristic in the reverse direction). Under reverse bias, filled states on the p-side become increasingly aligned with empty states on the n-side and electrons now tunnel through the pn junction barrier in reverse direction.

Technical Comparisons

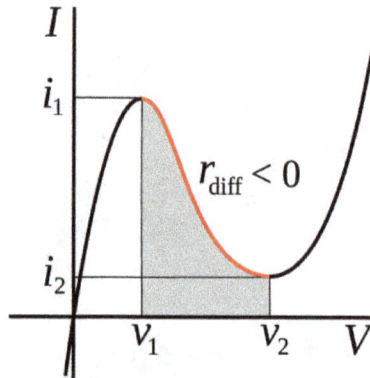

IV curve similar to a tunnel diode characteristic curve. It has negative differential resistance in the shaded voltage region, between v_1 and v_2.

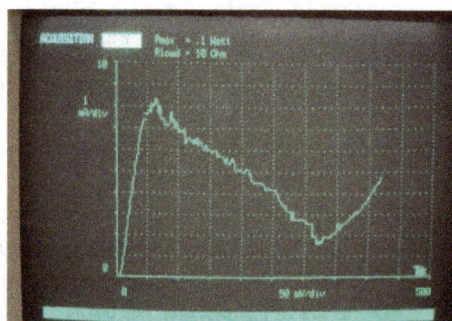

I-V curve of 10mA germanium tunnel diode, taken on a Tektronix model 571 curve tracer.

In a conventional semiconductor diode, conduction takes place while the p–n junction is forward biased and blocks current flow when the junction is reverse biased. This occurs up to a point known as the "reverse breakdown voltage" at which point conduction begins (often accompanied by destruction of the device). In the tunnel diode, the dopant concentrations in the p and n layers are increased to a level such that the reverse breakdown voltage becomes zero and the diode conducts in the reverse direction. However, when forward-biased, an effect occurs called quantum mechanical tunneling which gives rise to a region in its voltage-current behavior where an *increase* in forward voltage is accompanied by a *decrease* in forward current. This negative resistance region can be exploited in a solid state version of the dynatron oscillator which normally uses a tetrode thermionic valve (vacuum tube).

The tunnel diode showed great promise as an oscillator and high-frequency threshold (trigger) device since it operated at frequencies far greater than the tetrode could, well into the microwave bands. Applications for tunnel diodes included local oscillators for UHF television tuners, trigger circuits in oscilloscopes, high-speed counter circuits, and very fast-rise time pulse generator circuits. The tunnel diode can also be used as a low-noise microwave amplifier. However, since its discovery, more conventional semiconductor devices have surpassed its performance using conventional oscillator techniques. For many purposes, a three-terminal device, such as a field-effect transistor, is more flexible than a device with only two terminals. Practical tunnel diodes operate at a few milliamperes and a few tenths of a volt, making them low-power devices. The Gunn diode has similar high frequency capability and can handle more power.

Tunnel diodes are also more resistant to ionizing radiation than other diodes. This makes them well suited to higher radiation environments such as those found in space.

Longevity

Tunnel diodes are notable for their longevity, with devices made in the 1960s still functioning. Writing in *Nature*, Esaki and coauthors state that semiconductor devices in general are extremely stable, and suggest that their shelf life should be "infinite" if kept at room temperature. They go on to report that a small-scale test of 50-year-old devices revealed a "gratifying confirmation of the diode's longevity". As noticed on some samples of Esaki diodes, the gold plated iron pins can in fact corrode and short out to the case. This can usually be diagnosed, and the diode inside normally still works.

However, these components are susceptible to damage by overheating, and thus special care is needed when soldering them.

Thyristor

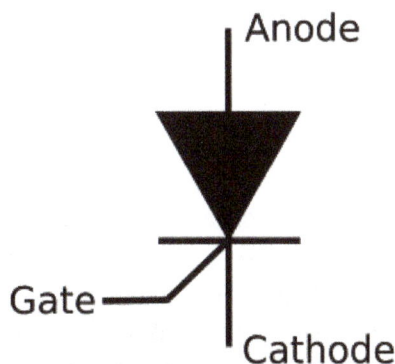

Circuit symbol for thyristor

A thyristor is a solid-state semiconductor device with four layers of alternating N and P-type material. It acts exclusively as a bistable switch, conducting when the gate receives a current trigger, and continuing to conduct while the voltage across the device is not reversed (forward-biased). A three-lead thyristor is designed to control the larger current of its two leads by combining that current with the smaller current of its other lead, known as its control lead. In contrast, a two-lead thyristor is designed to switch on if the potential difference between its leads is sufficiently large (breakdown voltage).

An SCR rated about 100 amperes, 1200 volts mounted on a heat sink - the two small wires are the gate trigger leads

Some sources define silicon-controlled rectifier (SCR) and thyristor as synonymous. Other sources define thyristors as a larger set of devices with at least four layers of alternating N and P-type material.

The first thyristor devices were released commercially in 1956. Because thyristors can control a relatively large amount of power and voltage with a small device, they find wide application in control of electric power, ranging from light dimmers and electric motor speed control to high-voltage direct current power transmission. Thyristors may be used in power-switching circuits, relay-replacement circuits, inverter circuits, oscillator circuits, level-detector circuits, chopper circuits, light-dimming circuits, low-cost timer circuits, logic circuits, speed-control circuits, phase-control circuits, etc. Originally, thyristors relied only on current reversal to turn them off, making them difficult to apply for direct current; newer device types can be turned on and off through the control gate signal. The latter is known as a gate turn-off thyristor, or GTO thyristor. A thyristor is not a proportional device like a transistor. In other words, a thyristor can only be fully on or off, while a transistor can lie in between on and off states. This makes a thyristor unsuitable as an analog amplifier, but useful as a switch.

Introduction

The thyristor is a four-layered, three terminal semiconductor device, with each layer consisting of alternately N-type or P-type material, for example P-N-P-N. The main terminals, labelled anode and cathode, are across all four layers. The control terminal, called the gate, is attached to p-type material near the cathode. (A variant called an SCS—Silicon Controlled Switch—brings all four layers out to terminals.) The operation of a thyristor can be understood in terms of a pair of tightly coupled bipolar junction transistors, arranged to cause a self-latching action:

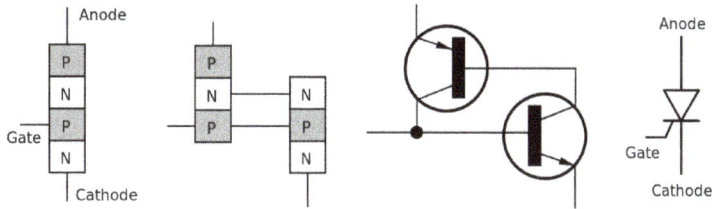

Structure on the physical and electronic level, and the thyristor symbol.

Thyristors have three states:

- Reverse blocking mode — Voltage is applied in the direction that would be blocked by a diode

- Forward blocking mode — Voltage is applied in the direction that would cause a diode to conduct, but the thyristor has not been triggered into conduction

- Forward conducting mode — The thyristor has been triggered into conduction and will remain conducting until the forward current drops below a threshold value known as the "holding current"

Function of The Gate Terminal

The thyristor has three p-n junctions (serially named J_1, J_2, J_3 from the anode).

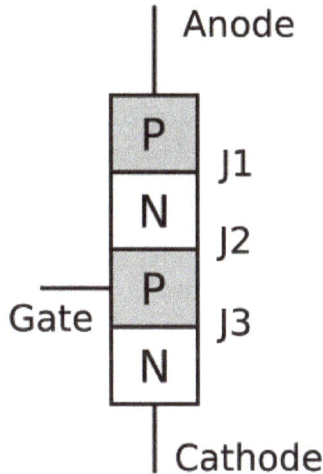

Layer diagram of thyristor.

When the anode is at a positive potential V_{AK} with respect to the cathode with no voltage applied at the gate, junctions J_1 and J_3 are forward biased, while junction J_2 is reverse biased. As J_2 is reverse biased, no conduction takes place (Off state). Now if V_{AK} is increased beyond the breakdown voltage V_{BO} of the thyristor, avalanche breakdown of J_2 takes place and the thyristor starts conducting (On state).

If a positive potential V_G is applied at the gate terminal with respect to the cathode, the breakdown of the junction J_2 occurs at a lower value of V_{AK}. By selecting an appropriate value of V_G, the thyristor can be switched into the on state quickly.

Once avalanche breakdown has occurred, the thyristor continues to conduct, irrespective of the gate voltage, until: (a) the potential V_{AK} is removed or (b) the current through the device (anode–cathode) is less than the holding current specified by the manufacturer. Hence V_G can be a voltage pulse, such as the voltage output from a UJT relaxation oscillator.

The gate pulses are characterized in terms of gate trigger voltage (V_{GT}) and gate trigger current (I_{GT}). Gate trigger current varies inversely with gate pulse width in such a way that it is evident that there is a minimum gate charge required to trigger the thyristor.

Switching Characteristics

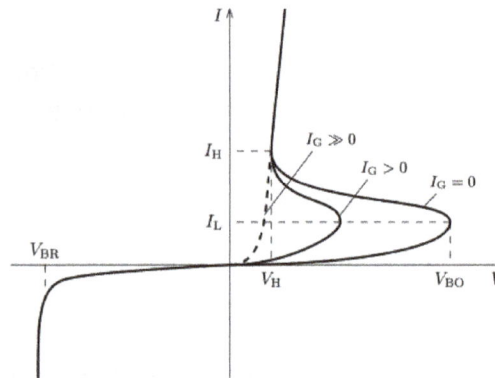

V - I characteristics.

In a conventional thyristor, once it has been switched on by the gate terminal, the device remains latched in the on-state (*i.e.* does not need a continuous supply of gate current to remain in the on state), providing the anode current has exceeded the latching current (I_L). As long as the anode remains positively biased, it cannot be switched off until the anode current falls below the holding current (I_H). In normal working condition the latching current is always greater than holding current. In the above figure I_L *has to come above the* I_H *on y-axis since* $I_L > I_H$.

A thyristor can be switched off if the external circuit causes the anode to become negatively biased (a method known as natural, or line, commutation). In some applications this is done by switching a second thyristor to discharge a capacitor into the cathode of the first thyristor. This method is called forced commutation.

After the current in a thyristor has extinguished, a finite time delay must elapse before the anode can again be positively biased *and* retain the thyristor in the off-state. This minimum delay is called the circuit commutated turn off time (t_Q). Attempting to positively bias the anode within this time causes the thyristor to be self-triggered by the remaining charge carriers (holes and electrons) that have not yet recombined.

For applications with frequencies higher than the domestic AC mains supply (e.g. 50 Hz or 60 Hz), thyristors with lower values of t_Q are required. Such fast thyristors can be made by diffusing heavy metal ions such as gold or platinum which act as charge combination centers into the silicon. Today, fast thyristors are more usually made by electron or proton irradiation of the silicon, or by ion implantation. Irradiation is more versatile than heavy metal doping because it permits the dosage to be adjusted in fine steps, even at quite a late stage in the processing of the silicon.

History

The silicon controlled rectifier (SCR) or thyristor proposed by William Shockley in 1950 and championed by Moll and others at Bell Labs was developed in 1956 by power engineers at General Electric (G.E.), led by Gordon Hall and commercialized by G.E.'s Frank W. "Bill" Gutzwiller.

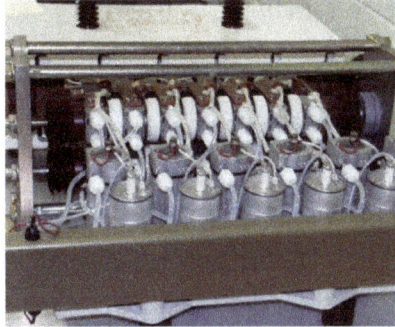

A bank of six 2000 A thyristors (white disks arranged in a row at top, and seen edge-on)

Etymology

An earlier gas filled tube device called a thyratron provided a similar electronic switching capability, where a small control voltage could switch a large current. It is from a combination of "thyratron" and "transistor" that the term "thyristor" is derived.

Applications

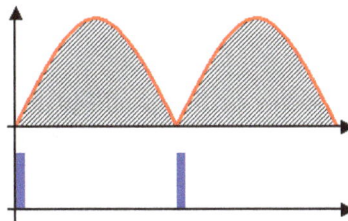

Waveforms in a thyristor circuit controlling an AC current.
Red trace: load (output) voltage
Blue trace: trigger voltage.

Thyristors are mainly used where high currents and voltages are involved, and are often used to control alternating currents, where the change of polarity of the current causes the device to switch off automatically, referred to as "zero cross" operation. The device can be said to operate *synchronously*; being that, once the device is triggered, it conducts current in phase with the voltage applied over its cathode to anode junction with no further gate modulation being required, i.e., the device is biased *fully on*. This is not to be confused with asymmetrical operation, as the output is unidirectional, flowing only from cathode to anode, and so is asymmetrical in nature.

Thyristors can be used as the control elements for phase angle triggered controllers, also known as phase fired controllers.

They can also be found in power supplies for digital circuits, where they are used as a sort of "enhanced circuit breaker" to prevent a failure in the power supply from damaging downstream components. A thyristor is used in conjunction with a Zener diode attached to its gate, and if the output voltage of the supply rises above the Zener voltage, the thyristor will conduct and short-circuit the power supply output to ground (in general also tripping an upstream breaker or fuse). This kind of protection circuit is known as a crowbar, and has the advantage over a standard circuit breaker or fuse in that it creates a high-conductance path to ground for the damaging supply voltage and potentially for stored energy in the system being powered.

The first large-scale application of thyristors, with associated triggering diac, in consumer products related to stabilized power supplies within color television receivers in the early 1970s. or The stabilized high voltage DC supply for the receiver was obtained by moving the switching point of the thyristor device up and down the falling slope of the positive going half of the AC supply input (if the rising slope was used the output voltage would always rise towards the peak input voltage when the device was triggered and thus defeat the aim of regulation). The precise switching point was determined by the load on the DC output supply, as well as AC input fluctuations.

Thyristors have been used for decades as lighting dimmers, in television, motion pictures, and theater, where they replaced inferior technologies such as autotransformers and rheostats. They have also been used in photography as a critical part of flashes (strobes).

Snubber Circuits

Thyristors can be triggered by a high rise-rate of off-state voltage. This is prevented by connecting a resistor-capacitor (RC) snubber circuit between the anode and cathode terminals in order to limit the dV/dt (i.e., rate of voltage change over time).

HVDC Electricity Transmission

Since modern thyristors can switch power on the scale of megawatts, thyristor valves have become the heart of high-voltage direct current (HVDC) conversion either to or from alternating current. In the realm of this and other very high-power applications, both electrically triggered (ETT) and light-triggered (LTT) thyristors are still the primary choice. The valves are arranged in stacks usually suspended from the ceiling of a transmission building called a valve hall. Thyristors are arranged into a diode bridge circuit and to reduce harmonics are connected in series to form a 12-pulse converter. Each thyristor is cooled with deionized water, and the entire arrangement becomes one of multiple identical modules forming a layer in a multilayer valve stack called a *quadruple valve*. Three such stacks are typically mounted on the floor or hung from the ceiling of the valve hall of a long-distance transmission facility.

Valve hall containing thyristor valve stacks used for long-distance transmission of power from Manitoba Hydro dams

Comparisons to Other Devices

The functional drawback of a thyristor is that, like a diode, it only conducts in one direction. A similar self-latching 5-layer device, called a TRIAC, is able to work in both directions. This added capability, though, also can become a shortfall. Because the TRIAC can conduct in both directions, reactive loads can cause it to fail to turn off during the zero-voltage instants of the AC power cycle. Because of this, use of TRIACs with (for example) heavily inductive motor loads usually requires the use of a "snubber" circuit around the TRIAC to assure that it will turn off with each half-cycle of mains power. Inverse parallel SCRs can also be used in place of the triac; because each SCR in the pair has an entire half-cycle of reverse polarity applied to it, the SCRs, unlike TRIACs, are sure to turn off. The "price" to be paid for this arrangement, however, is the added complexity of two separate, but essentially identical gating circuits.

Although thyristors are heavily used in megawatt-scale rectification of AC to DC, in low- and medium-power (from few tens of watts to few tens of kilowatts) applications they have virtually been replaced by other devices with superior switching characteristics like Power MOSFETs or IGBTs. One major problem associated with SCRs is that they are not fully controllable switches. The GTO thyristor and IGCT are two devices related to the thyristor that address this problem. In high-frequency applications, thyristors are poor candidates due to large switching times arising from bipolar conduction. MOSFETs, on the other hand, have much faster switching capability because of their unipolar conduction (only majority carriers carry the current).

Failure Modes

Thyristor manufacturers generally specify a region of safe firing defining acceptable levels of voltage and current for a given operating temperature. The boundary of this region is partly determined by the requirement that the maximum permissible gate power (P_G), specified for a given trigger pulse duration, is not exceeded.

As well as the usual failure modes due to exceeding voltage, current or power ratings, thyristors have their own particular modes of failure, including:

- Turn on di/dt — in which the rate of rise of on-state current after triggering is higher than can be supported by the spreading speed of the active conduction area (SCRs & triacs).

- Forced commutation — in which the transient peak reverse recovery current causes such a high voltage drop in the sub-cathode region that it exceeds the reverse breakdown voltage of the gate cathode diode junction (SCRs only).

- Switch on dv/dt — the thyristor can be spuriously fired without trigger from the gate if the anode-to-cathode voltage rise-rate is too great.

Silicon Carbide Thyristors

In recent years, some manufacturers have developed thyristors using silicon carbide (SiC) as the semiconductor material. These have applications in high temperature environments, being capable of operating at temperatures up to 350 °C.

Types

- ACS
- ACST
- AGT — Anode Gate Thyristor — A thyristor with gate on n-type layer near to the anode
- ASCR — Asymmetrical SCR
- BCT — Bidirectional Control Thyristor — A bidirectional switching device containing two thyristor structures with separate gate contacts
- BOD — Breakover Diode — A gateless thyristor triggered by avalanche current
  - DIAC — Bidirectional trigger device
  - Dynistor — Unidirectional switching device
  - Shockley diode — Unidirectional trigger and switching device
  - SIDAC — Bidirectional switching device
  - Trisil, SIDACtor — Bidirectional protection devices
- BRT — Base Resistance Controlled Thyristor
- ETO — Emitter Turn-Off Thyristor
- GTO — Gate Turn-Off thyristor
  - DB-GTO — Distributed buffer gate turn-off thyristor
  - MA-GTO — Modified anode gate turn-off thyristor
- IGCT — Integrated gate-commutated thyristor
- Ignitor — Spark generators for fire-lighter ckts
- LASCR — Light-activated SCR, or LTT — light-triggered thyristor
- LASS — light-activated semiconducting switch
- MCT — MOSFET Controlled Thyristor — It contains two additional FET structures for on/off control.
- CSMT or MCS — MOS composite static induction thyristor
- PUT or PUJT — Programmable Unijunction Transistor — A thyristor with gate on n-type layer near to the anode used as a functional replacement for unijunction transistor
- RCT — Reverse Conducting Thyristor
- SCS — Silicon Controlled Switch or Thyristor Tetrode — A thyristor with both cathode and anode gates
- SCR — Silicon Controlled Rectifier

- SITh — Static Induction Thyristor, or FCTh — Field Controlled Thyristor — containing a gate structure that can shut down anode current flow.

- TRIAC — Triode for Alternating Current — A bidirectional switching device containing two thyristor structures with common gate contact

- Quadrac — special type of thyristor which combines a DIAC and a TRIAC into a single package.

Reverse Conducting Thyristor

A reverse conducting thyristor (RCT) has an integrated reverse diode, so is not capable of reverse blocking. These devices are advantageous where a reverse or freewheel diode must be used. Because the SCR and diode never conduct at the same time they do not produce heat simultaneously and can easily be integrated and cooled together. Reverse conducting thyristors are often used in frequency changers and inverters.

Photothyristors

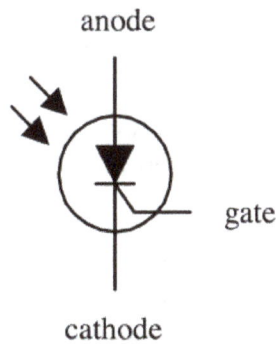

Electronic symbol for light-activated SCR (LASCR)

Photothyristors are activated by light. The advantage of photothyristors is their insensitivity to electrical signals, which can cause faulty operation in electrically noisy environments. A light-triggered thyristor (LTT) has an optically sensitive region in its gate, into which electromagnetic radiation (usually infrared) is coupled by an optical fiber. Since no electronic boards need to be provided at the potential of the thyristor in order to trigger it, light-triggered thyristors can be an advantage in high-voltage applications such as HVDC. Light-triggered thyristors are available with in-built over-voltage (VBO) protection, which triggers the thyristor when the forward voltage across it becomes too high; they have also been made with in-built *forward recovery protection*, but not commercially. Despite the simplification they can bring to the electronics of an HVDC valve, light-triggered thyristors may still require some simple monitoring electronics and are only available from a few manufacturers.

Two common photothyristors include the light-activated SCR (LASCR) and the light-activated TRIAC. A LASCR acts as a switch that turns on when exposed to light. Following light exposure, when light is absent, if the power is not removed and the polarities of the cathode and anode have not yet reversed, the LASCR is still in the "on" state. A light-activated TRIAC resembles a LASCR, except that it is designed for alternating currents.

Transistor

Assorted discrete transistors. Packages in order from top to bottom: TO-3, TO-126, TO-92, SOT-23.

A transistor is a semiconductor device used to amplify or switch electronic signals and electrical power. It is composed of semiconductor material usually with at least three terminals for connection to an external circuit. A voltage or current applied to one pair of the transistor's terminals changes the current through another pair of terminals. Because the controlled (output) power can be higher than the controlling (input) power, a transistor can amplify a signal. Today, some transistors are packaged individually, but many more are found embedded in integrated circuits.

The transistor is the fundamental building block of modern electronic devices, and is ubiquitous in modern electronic systems. First conceived by Julius Lilienfeld in 1926 and practically implemented in 1947 by American physicists John Bardeen, Walter Brattain, and William Shockley, the transistor revolutionized the field of electronics, and paved the way for smaller and cheaper radios, calculators, and computers, among other things. The transistor is on the list of IEEE milestones in electronics, and Bardeen, Brattain, and Shockley shared the 1956 Nobel Prize in Physics for their achievement.

History

A replica of the first working transistor.

The thermionic triode, a vacuum tube invented in 1907, enabled amplified radio technology and long-distance telephony. The triode, however, was a fragile device that consumed a lot of power.

Physicist Julius Edgar Lilienfeld filed a patent for a field-effect transistor (FET) in Canada in 1925, which was intended to be a solid-state replacement for the triode. Lilienfeld also filed identical patents in the United States in 1926 and 1928. However, Lilienfeld did not publish any research articles about his devices nor did his patents cite any specific examples of a working prototype. Because the production of high-quality semiconductor materials was still decades away, Lilienfeld's solid-state amplifier ideas would not have found practical use in the 1920s and 1930s, even if such a device had been built. In 1934, German inventor Oskar Heil patented a similar device.

John Bardeen, William Shockley and Walter Brattain at Bell Labs, 1948.

From November 17, 1947 to December 23, 1947, John Bardeen and Walter Brattain at AT&T's Bell Labs in the United States performed experiments and observed that when two gold point contacts were applied to a crystal of germanium, a signal was produced with the output power greater than the input. Solid State Physics Group leader William Shockley saw the potential in this, and over the next few months worked to greatly expand the knowledge of semiconductors. The term *transistor* was coined by John R. Pierce as a contraction of the term *transresistance*. According to Lillian Hoddeson and Vicki Daitch, authors of a biography of John Bardeen, Shockley had proposed that Bell Labs' first patent for a transistor should be based on the field-effect and that he be named as the inventor. Having unearthed Lilienfeld's patents that went into obscurity years earlier, lawyers at Bell Labs advised against Shockley's proposal because the idea of a field-effect transistor that used an electric field as a "grid" was not new. Instead, what Bardeen, Brattain, and Shockley invented in 1947 was the first point-contact transistor. In acknowledgement of this accomplishment, Shockley, Bardeen, and Brattain were jointly awarded the 1956 Nobel Prize in Physics "for their researches on semiconductors and their discovery of the transistor effect."

In 1948, the point-contact transistor was independently invented by German physicists Herbert Mataré and Heinrich Welker while working at the Compagnie des Freins et Signaux, a Westinghouse subsidiary located in Paris. Mataré had previous experience in developing crystal rectifiers from silicon and germanium in the German radar effort during World War II. Using this knowledge, he began researching the phenomenon of "interference" in 1947. By June 1948, witnessing currents flowing through point-contacts, Mataré produced consistent results using samples of germanium produced by Welker, similar to what Bardeen and Brattain had accomplished earlier in December 1947. Realizing that Bell Labs' scientists had already invented the transistor before them, the company rushed to get its "transistron" into production for amplified use in France's telephone network.

Herbert F. Mataré (1950)

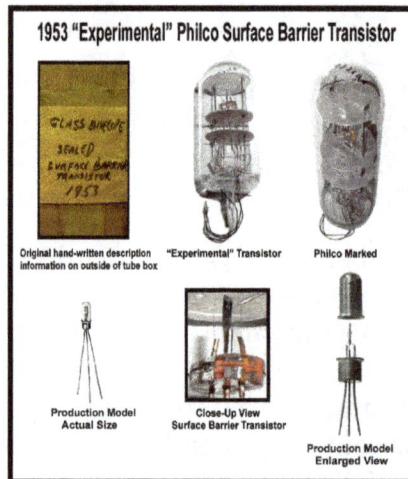

Philco surface-barrier transistor developed and produced in 1953

The first high-frequency transistor was the surface-barrier germanium transistor developed by Philco in 1953, capable of operating up to 60 MHz. These were made by etching depressions into an N-type germanium base from both sides with jets of Indium(III) sulfate until it was a few ten-thousandths of an inch thick. Indium electroplated into the depressions formed the collector and emitter.

The first "prototype" pocket transistor radio was shown by INTERMETALL (a company founded by Herbert Mataré in 1952) at the Internationale Funkausstellung Düsseldorf between August 29, 1953 and September 9, 1953.

The first "production" all-transistor car radio was produced in 1955 by Chrysler and Philco, had used surface-barrier transistors in its circuitry and which were also first suitable for high-speed computers.

The first working silicon transistor was developed at Bell Labs on January 26, 1954 by Morris Tanenbaum. The first commercial silicon transistor was produced by Texas Instruments in 1954. This was the work of Gordon Teal, an expert in growing crystals of high purity, who had previously worked at Bell Labs. The first MOS transistor actually built was by Kahng and Atalla at Bell Labs in 1960.

Importance

The transistor is the key active component in practically all modern electronics. Many consider it to be one of the greatest inventions of the 20th century. Its importance in today's society rests on its ability to be mass-produced using a highly automated process (semiconductor device fabrication) that achieves astonishingly low per-transistor costs. The invention of the first transistor at Bell Labs was named an IEEE Milestone in 2009.

A Darlington transistor opened up so the actual transistor chip (the small square) can be seen inside. A Darlington transistor is effectively two transistors on the same chip. One transistor is much larger than the other, but both are large in comparison to transistors in large-scale integration because this particular example is intended for power applications.

Although several companies each produce over a billion individually packaged (known as *discrete*) transistors every year, the vast majority of transistors are now produced in integrated circuits (often shortened to *IC*, *microchips* or simply *chips*), along with diodes, resistors, capacitors and other electronic components, to produce complete electronic circuits. A logic gate consists of up to about twenty transistors whereas an advanced microprocessor, as of 2009, can use as many as 3 billion transistors (MOSFETs). "About 60 million transistors were built in 2002... for [each] man, woman, and child on Earth."

The transistor's low cost, flexibility, and reliability have made it a ubiquitous device. Transistorized mechatronic circuits have replaced electromechanical devices in controlling appliances and machinery. It is often easier and cheaper to use a standard microcontroller and write a computer program to carry out a control function than to design an equivalent mechanical control function.

Simplified Operation

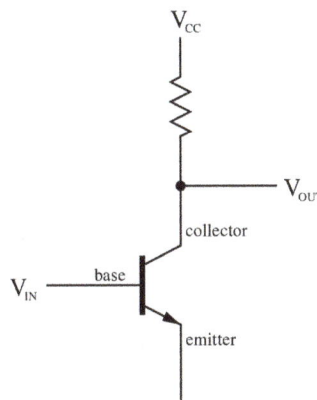

A simple circuit diagram to show the labels of a n–p–n bipolar transistor.

The essential usefulness of a transistor comes from its ability to use a small signal applied between one pair of its terminals to control a much larger signal at another pair of terminals. This property is called gain. It can produce a stronger output signal, a voltage or current, which is proportional to a weaker input signal; that is, it can act as an amplifier. Alternatively, the transistor can be used to turn current on or off in a circuit as an electrically controlled switch, where the amount of current is determined by other circuit elements.

There are two types of transistors, which have slight differences in how they are used in a circuit. A *bipolar transistor* has terminals labeled base, collector, and emitter. A small current at the base terminal (that is, flowing between the base and the emitter) can control or switch a much larger current between the collector and emitter terminals. For a *field-effect transistor*, the terminals are labeled gate, source, and drain, and a voltage at the gate can control a current between source and drain.

The image represents a typical bipolar transistor in a circuit. Charge will flow between emitter and collector terminals depending on the current in the base. Because internally the base and emitter connections behave like a semiconductor diode, a voltage drop develops between base and emitter while the base current exists. The amount of this voltage depends on the material the transistor is made from, and is referred to as V_{BE}.

Transistor as a Switch

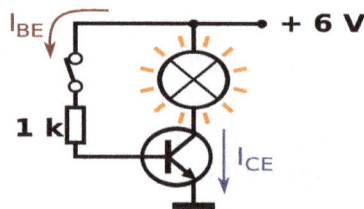

BJT used as an electronic switch, in grounded-emitter configuration.

Transistors are commonly used in digital circuits as electronic switches which can be either in an "on" or "off" state, both for high-power applications such as switched-mode power supplies and for low-power applications such as logic gates. Important parameters for this application include the current switched, the voltage handled, and the switching speed, characterised by the rise and fall times.

In a grounded-emitter transistor circuit, such as the light-switch circuit shown, as the base voltage rises, the emitter and collector currents rise exponentially. The collector voltage drops because of reduced resistance from collector to emitter. If the voltage difference between the collector and emitter were zero (or near zero), the collector current would be limited only by the load resistance (light bulb) and the supply voltage. This is called *saturation* because current is flowing from collector to emitter freely. When saturated, the switch is said to be *on*.

Providing sufficient base drive current is a key problem in the use of bipolar transistors as switches. The transistor provides current gain, allowing a relatively large current in the collector to be switched by a much smaller current into the base terminal. The ratio of these currents varies depending on the type of transistor, and even for a particular type, varies depending on the collector current. In the example light-switch circuit shown, the resistor is chosen to provide enough base current to ensure the transistor will be saturated.

In a switching circuit, the idea is to simulate, as near as possible, the ideal switch having the properties of open circuit when off, short circuit when on, and an instantaneous transition between the two states. Parameters are chosen such that the "off" output is limited to leakage currents too small to affect connected circuitry; the resistance of the transistor in the "on" state is too small to affect circuitry; and the transition between the two states is fast enough not to have a detrimental effect.

Transistor as an Amplifier

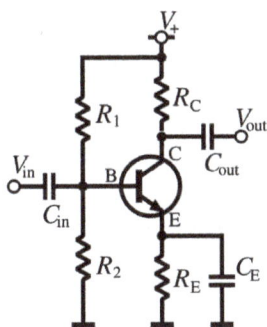

Amplifier circuit, common-emitter configuration with a voltage-divider bias circuit.

The common-emitter amplifier is designed so that a small change in voltage (V_{in}) changes the small current through the base of the transistor; the transistor's current amplification combined with the properties of the circuit mean that small swings in V_{in} produce large changes in V_{out}.

Various configurations of single transistor amplifier are possible, with some providing current gain, some voltage gain, and some both.

From mobile phones to televisions, vast numbers of products include amplifiers for sound reproduction, radio transmission, and signal processing. The first discrete-transistor audio amplifiers barely supplied a few hundred milliwatts, but power and audio fidelity gradually increased as better transistors became available and amplifier architecture evolved.

Modern transistor audio amplifiers of up to a few hundred watts are common and relatively inexpensive.

Comparison with Vacuum Tubes

Before transistors were developed, vacuum (electron) tubes (or in the UK "thermionic valves" or just "valves") were the main active components in electronic equipment.

Advantages

The key advantages that have allowed transistors to replace vacuum tubes in most applications are

- no cathode heater (which produces the characteristic orange glow of tubes), reducing power consumption, eliminating delay as tube heaters warm up, and immune from cathode poisoning and depletion;

- very small size and weight, reducing equipment size;

- large numbers of extremely small transistors can be manufactured as a single integrated circuit;

- low operating voltages compatible with batteries of only a few cells;

- circuits with greater energy efficiency are usually possible. For low-power applications (e.g., voltage amplification) in particular, energy consumption can be very much less than for tubes;

- inherent reliability and very long life; tubes always degrade and fail over time. Some transistorized devices have been in service for more than 50 years ;

- complementary devices available, providing design flexibility including complementary-symmetry circuits, not possible with vacuum tubes;

- very low sensitivity to mechanical shock and vibration, providing physical ruggedness and virtually eliminating shock-induced spurious signals (e.g., microphonics in audio applications);

- not susceptible to breakage of a glass envelope, leakage, outgassing, and other physical damage.

Limitations

Transistors have the following limitations:

- silicon transistors can age and fail;

- high-power, high-frequency operation, such as that used in over-the-air television broadcasting, is better achieved in vacuum tubes due to improved electron mobility in a vacuum;

- solid-state devices are susceptible to damage from very brief electrical and thermal events, including electrostatic discharge in handling; vacuum tubes are electrically much more rugged;

- sensitivity to radiation and cosmic rays (special radiation-hardened chips are used for spacecraft devices);

- vacuum tubes in audio applications create significant lower-harmonic distortion, the so-called tube sound, which some people prefer.

Types

BJT and JFET symbols

JFET	MOSFET enh		MOSFET dep	

(top row) P-channel
(bottom row) N-channel

JFET and MOSFET symbols

Transistors are categorized by

- semiconductor material: the metalloids germanium (first used in 1947) and silicon (first used in 1954)—in amorphous, polycrystalline and monocrystalline form—, the compounds gallium arsenide (1966) and silicon carbide (1997), the alloy silicon-germanium (1989), the allotrope of carbon graphene (research ongoing since 2004), etc.;

- structure: BJT, JFET, IGFET (MOSFET), insulated-gate bipolar transistor, "other types";

- electrical polarity (positive and negative): n–p–n, p–n–p (BJTs), n-channel, p-channel (FETs);

- maximum power rating: low, medium, high;

- maximum operating frequency: low, medium, high, radio (RF), microwave frequency (the maximum effective frequency of a transistor in a common-emitter or common-source circuit is denoted by the term f_T, an abbreviation for transition frequency—the frequency of transition is the frequency at which the transistor yields unity voltage gain)

- application: switch, general purpose, audio, high voltage, super-beta, matched pair;

- physical packaging: through-hole metal, through-hole plastic, surface mount, ball grid array, power modules;

- amplification factor h_{FE}, β_F (transistor beta) or g_m (transconductance).

Hence, a particular transistor may be described as *silicon, surface-mount, BJT, n–p–n, low-power, high-frequency switch.*

A popular way to remember which symbol represents which type of transistor is to look at the arrow and how it is arranged. Within an NPN transistor symbol, the arrow will Not Point iN. Conversely, within the PNP symbol you see that the arrow Points iN Proudly.

Bipolar Junction Transistor (BJT)

Bipolar transistors are so named because they conduct by using both majority and minority carriers. The bipolar junction transistor, the first type of transistor to be mass-produced, is a combination of two junction diodes, and is formed of either a thin layer of p-type semiconductor sandwiched between two n-type semiconductors (an n–p–n transistor), or a thin layer of n-type semiconductor sandwiched between two p-type semiconductors (a p–n–p transistor). This construction produces two p–n junctions: a base–emitter junction and a base–collector junction, separated by a thin region of semiconductor known as the base region (two junction diodes wired together without sharing an intervening semiconducting region will not make a transistor).

BJTs have three terminals, corresponding to the three layers of semiconductor—an *emitter*, a *base*, and a *collector*. They are useful in amplifiers because the currents at the emitter and collector are controllable by a relatively small base current. In an n–p–n transistor operating in the active region, the emitter–base junction is forward biased (electrons and holes recombine at the junction), and electrons are injected into the base region. Because the base is narrow, most of these electrons will diffuse into the reverse-biased (electrons and holes are formed at, and move away from the junction) base–collector junction and be swept into the collector; perhaps one-hundredth of the electrons will recombine in the base, which is the dominant mechanism in the base current. By controlling the number of electrons that can leave the base, the number of electrons entering the collector can be controlled. Collector current is approximately β (common-emitter current gain) times the base current. It is typically greater than 100 for small-signal transistors but can be smaller in transistors designed for high-power applications.

Unlike the field-effect transistor, the BJT is a low-input-impedance device. Also, as the base–emitter voltage (V_{BE}) is increased the base–emitter current and hence the collector–emitter current (I_{CE}) increase exponentially according to the Shockley diode model and the Ebers-Moll model. Because of this exponential relationship, the BJT has a higher transconductance than the FET.

Bipolar transistors can be made to conduct by exposure to light, because absorption of photons in the base region generates a photocurrent that acts as a base current; the collector current is approximately β times the photocurrent. Devices designed for this purpose have a transparent window in the package and are called phototransistors.

Field-Effect Transistor (FET)

The *field-effect transistor*, sometimes called a *unipolar transistor*, uses either electrons (in *n-channel FET*) or holes (in *p-channel FET*) for conduction. The four terminals of the FET are named *source*, *gate*, *drain*, and *body* (*substrate*). On most FETs, the body is connected to the source inside the package, and this will be assumed for the following description.

In a FET, the drain-to-source current flows via a conducting channel that connects the *source* region to the *drain* region. The conductivity is varied by the electric field that is produced when a voltage is applied between the gate and source terminals; hence the current flowing between the drain and source is controlled by the voltage applied between the gate and source. As the gate–source voltage (V_{GS}) is increased, the drain–source current (I_{DS}) increases exponentially for V_{GS} below threshold, and then at a roughly quadratic rate ($I_{GS} \propto (V_{GS} - V_{T})^2$) (where V_{T} is the threshold voltage at which drain current begins) in the "space-charge-limited" region above threshold. A quadratic behavior is not observed in modern devices, for example, at the 65 nm technology node.

For low noise at narrow bandwidth the higher input resistance of the FET is advantageous.

FETs are divided into two families: *junction FET* (JFET) and *insulated gate FET* (IGFET). The IGFET is more commonly known as a *metal–oxide–semiconductor FET* (MOSFET), reflecting its original construction from layers of metal (the gate), oxide (the insulation), and semiconductor. Unlike IGFETs, the JFET gate forms a p–n diode with the channel which lies between the source and drain. Functionally, this makes the n-channel JFET the solid-state equivalent of the vacuum tube triode which, similarly, forms a diode between its grid and cathode. Also, both devices operate in the *depletion mode*, they both have a high input impedance, and they both conduct current under the control of an input voltage.

Metal–semiconductor FETs (MESFETs) are JFETs in which the reverse biased p–n junction is replaced by a metal–semiconductor junction. These, and the HEMTs (high-electron-mobility transistors, or HFETs), in which a two-dimensional electron gas with very high carrier mobility is used for charge transport, are especially suitable for use at very high frequencies (microwave frequencies; several GHz).

FETs are further divided into *depletion-mode* and *enhancement-mode* types, depending on whether the channel is turned on or off with zero gate-to-source voltage. For enhancement mode, the channel is off at zero bias, and a gate potential can "enhance" the conduction. For the depletion mode, the channel is on at zero bias, and a gate potential (of the opposite polarity) can "deplete" the channel, reducing conduction. For either mode, a more positive gate voltage corresponds to a higher current for n-channel devices and a lower current for p-channel devices. Nearly all JFETs are depletion-mode because the diode junctions would forward bias and conduct if they were enhancement-mode devices; most IGFETs are enhancement-mode types.

Usage of Bipolar and Field-Effect Transistors

The bipolar junction transistor (BJT) was the most commonly used transistor in the 1960s and 70s. Even after MOSFETs became widely available, the BJT remained the transistor of choice for many analog circuits such as amplifiers because of their greater linearity and ease of manufacture. In integrated circuits, the desirable properties of MOSFETs allowed them to capture nearly all market share for digital circuits. Discrete MOSFETs can be applied in transistor applications, including analog circuits, voltage regulators, amplifiers, power transmitters and motor drivers.

Other Transistor Types

Transistor symbol created on Portuguese pavement in the University of Aveiro.

- Bipolar junction transistor (BJT):

    - heterojunction bipolar transistor, up to several hundred GHz, common in modern ultrafast and RF circuits;

    - Schottky transistor;

    - avalanche transistor:

    - Darlington transistors are two BJTs connected together to provide a high current gain equal to the product of the current gains of the two transistors;

    - insulated-gate bipolar transistors (IGBTs) use a medium-power IGFET, similarly connected to a power BJT, to give a high input impedance. Power diodes are often connected between certain terminals depending on specific use. IGBTs are particularly suitable for heavy-duty industrial applications. The Asea Brown Boveri (ABB) *5SNA2400E170100* illustrates just how far power semiconductor technology has advanced. Intended for three-phase power supplies, this device houses three n–p–n IGBTs in a case measuring 38 by 140 by 190 mm and weighing 1.5 kg. Each IGBT is rated at 1,700 volts and can handle 2,400 amperes;

    - phototransistor;

    - multiple-emitter transistor, used in transistor–transistor logic and integrated current mirrors;

    - multiple-base transistor, used to amplify very-low-level signals in noisy environments such as the pickup of a record player or radio front ends. Effectively, it is a very large number of transistors in parallel where, at the output, the signal is added constructively, but random noise is added only stochastically.

- Field-effect transistor (FET):

    - carbon nanotube field-effect transistor (CNFET), where the channel material is replaced by a carbon nanotube;

    - junction gate field-effect transistor (JFET), where the gate is insulated by a reverse-biased p–n junction;

    - metal–semiconductor field-effect transistor (MESFET), similar to JFET with a Schottky junction instead of a p–n junction;

    - high-electron-mobility transistor (HEMT);

    - metal–oxide–semiconductor field-effect transistor (MOSFET), where the gate is insulated by a shallow layer of insulator;

    - inverted-T field-effect transistor (ITFET);

    - fin field-effect transistor (FinFET), source/drain region shapes fins on the silicon surface;

- fast-reverse epitaxial diode field-effect transistor (FREDFET);

- thin-film transistor, in LCDs;

- organic field-effect transistor (OFET), in which the semiconductor is an organic compound;

- ballistic transistor;

- floating-gate transistor, for non-volatile storage;

- FETs used to sense environment;

  - ion-sensitive field-effect transistor (IFSET), to measure ion concentrations in solution,

  - electrolyte–oxide–semiconductor field-effect transistor (EOSFET), neuro-chip,

  - deoxyribonucleic acid field-effect transistor (DNAFET).

- Tunnel field-effect transistor, where it switches by modulating quantum tunnelling through a barrier.

- Diffusion transistor, formed by diffusing dopants into semiconductor substrate; can be both BJT and FET.

- Unijunction transistor, can be used as simple pulse generators. It comprise a main body of either P-type or N-type semiconductor with ohmic contacts at each end (terminals *Base1* and *Base2*). A junction with the opposite semiconductor type is formed at a point along the length of the body for the third terminal (*Emitter*).

- Single-electron transistors (SET), consist of a gate island between two tunneling junctions. The tunneling current is controlled by a voltage applied to the gate through a capacitor.

- Nanofluidic transistor, controls the movement of ions through sub-microscopic, water-filled channels.

- Multigate devices:

  - tetrode transistor;

  - pentode transistor;

  - trigate transistor (prototype by Intel);

  - dual-gate field-effect transistors have a single channel with two gates in cascode; a configuration optimized for *high-frequency amplifiers*, *mixers*, and *oscillators*.

- Junctionless nanowire transistor (JNT), uses a simple nanowire of silicon surrounded by an electrically isolated "wedding ring" that acts to gate the flow of electrons through the wire.

- Vacuum-channel transistor, when in 2012, NASA and the National Nanofab Center in South Korea were reported to have built a prototype vacuum-channel transistor in only 150 nanometers in size, can be manufactured cheaply using standard silicon semiconductor processing, can operate at high speeds even in hostile environments, and could consume just as much power as a standard transistor.

- Organic electrochemical transistor.

Part Numbering Standards/Specifications

The types of some transistors can be parsed from the part number. There are three major semiconductor naming standards; in each the alphanumeric prefix provides clues to type of the device.

Japanese Industrial Standard (JIS)

JIS Transistor Prefix Table	
Prefix	Type of transistor
2SA	high-frequency p–n–p BJTs
2SB	audio-frequency p–n–p BJTs
2SC	high-frequency n–p–n BJTs
2SD	audio-frequency n–p–n BJTs
2SJ	P-channel FETs (both JFETs and MOSFETs)
2SK	N-channel FETs (both JFETs and MOSFETs)

The *JIS-C-7012* specification for transistor part numbers starts with "2S", e.g. 2SD965, but sometimes the "2S" prefix is not marked on the package – a 2SD965 might only be marked "D965"; a 2SC1815 might be listed by a supplier as simply "C1815". This series sometimes has suffixes (such as "R", "O", "BL", standing for "red", "orange", "blue", etc.) to denote variants, such as tighter h_{FE} (gain) groupings.

European Electronic Component Manufacturers Association (EECA)

The Pro Electron standard, the European Electronic Component Manufacturers Association part numbering scheme, begins with two letters: the first gives the semiconductor type (A for germanium, B for silicon, and C for materials like GaAs); the second letter denotes the intended use (A for diode, C for general-purpose transistor, etc.). A 3-digit sequence number (or one letter then 2 digits, for industrial types) follows. With early devices this indicated the case type. Suffixes may be used, with a letter (e.g. "C" often means high h_{FE}, such as in: BC549C) or other codes may follow to show gain (e.g. BC327-25) or voltage rating (e.g. BUK854-800A). The more common prefixes are:

Pro Electron / EECA Transistor Prefix Table				
Prefix class	Type and usage	Example	Equivalent	Reference
AC	Germanium small-signal AF transistor	AC126	NTE102A	Datasheet
AD	Germanium AF power transistor	AD133	NTE179	Datasheet
AF	Germanium small-signal RF transistor	AF117	NTE160	Datasheet

AL	Germanium RF power transistor	ALZ10	NTE100	Datasheet
AS	Germanium switching transistor	ASY28	NTE101	Datasheet
AU	Germanium power switching transistor	AU103	NTE127	Datasheet
BC	Silicon, small-signal transistor ("general purpose")	BC548	2N3904	Datasheet
BD	Silicon, power transistor	BD139	NTE375	Datasheet
BF	Silicon, RF (high frequency) BJT or FET	BF245	NTE133	Datasheet
BS	Silicon, switching transistor (BJT or MOSFET)	BS170	2N7000	Datasheet
BL	Silicon, high frequency, high power (for transmitters)	BLW60	NTE325	Datasheet
BU	Silicon, high voltage (for CRT horizontal deflection circuits)	BU2520A	NTE2354	Datasheet
CF	Gallium Arsenide small-signal Microwave transistor (MESFET)	CF739	—	Datasheet
CL	Gallium Arsenide Microwave power transistor (FET)	CLY10	—	Datasheet

Joint Electron Devices Engineering Council (JEDEC)

The JEDEC *EIA370* transistor device numbers usually start with "2N", indicating a three-terminal device (dual-gate field-effect transistors are four-terminal devices, so begin with 3N), then a 2, 3 or 4-digit sequential number with no significance as to device properties (although early devices with low numbers tend to be germanium). For example, 2N3055 is a silicon n–p–n power transistor, 2N1301 is a p–n–p germanium switching transistor. A letter suffix (such as "A") is sometimes used to indicate a newer variant, but rarely gain groupings.

Proprietary

Manufacturers of devices may have their own proprietary numbering system, for example CK722. Since devices are second-sourced, a manufacturer's prefix (like "MPF" in MPF102, which originally would denote a Motorola FET) now is an unreliable indicator of who made the device. Some proprietary naming schemes adopt parts of other naming schemes, for example a PN2222A is a (possibly Fairchild Semiconductor) 2N2222A in a plastic case (but a PN108 is a plastic version of a BC108, not a 2N108, while the PN100 is unrelated to other xx100 devices).

Military part numbers sometimes are assigned their own codes, such as the British Military CV Naming System.

Manufacturers buying large numbers of similar parts may have them supplied with "house numbers", identifying a particular purchasing specification and not necessarily a device with a standardized registered number. For example, an HP part 1854,0053 is a (JEDEC) 2N2218 transistor which is also assigned the CV number: CV7763

Naming Problems

With so many independent naming schemes, and the abbreviation of part numbers when printed on the devices, ambiguity sometimes occurs. For example, two different devices may be marked "J176" (one the J176 low-power JFET, the other the higher-powered MOSFET 2SJ176).

As older "through-hole" transistors are given surface-mount packaged counterparts, they tend to be assigned many different part numbers because manufacturers have their own systems to cope with the variety in pinout arrangements and options for dual or matched n–p–n+p–n–p devices in one pack. So even when the original device (such as a 2N3904) may have been assigned by a standards authority, and well known by engineers over the years, the new versions are far from standardized in their naming.

Construction

Semiconductor Material

Semiconductor material characteristics				
Semiconductor material	Junction forward voltage V @ 25 °C	Electron mobility m²/(V·s) @ 25 °C	Hole mobility m²/(V·s) @ 25 °C	Max. junction temp. °C
Ge	0.27	0.39	0.19	70 to 100
Si	0.71	0.14	0.05	150 to 200
GaAs	1.03	0.85	0.05	150 to 200
Al-Si junction	0.3	—	—	150 to 200

The first BJTs were made from germanium (Ge). Silicon (Si) types currently predominate but certain advanced microwave and high-performance versions now employ the *compound semiconductor* material gallium arsenide (GaAs) and the *semiconductor alloy* silicon germanium (SiGe). Single element semiconductor material (Ge and Si) is described as *elemental*.

Rough parameters for the most common semiconductor materials used to make transistors are given in the table to the right; these parameters will vary with increase in temperature, electric field, impurity level, strain, and sundry other factors.

The *junction forward voltage* is the voltage applied to the emitter–base junction of a BJT in order to make the base conduct a specified current. The current increases exponentially as the junction forward voltage is increased. The values given in the table are typical for a current of 1 mA (the same values apply to semiconductor diodes). The lower the junction forward voltage the better, as this means that less power is required to "drive" the transistor. The junction forward voltage for a given current decreases with increase in temperature. For a typical silicon junction the change is −2.1 mV/°C. In some circuits special compensating elements (sensistors) must be used to compensate for such changes.

The density of mobile carriers in the channel of a MOSFET is a function of the electric field forming the channel and of various other phenomena such as the impurity level in the channel. Some impurities, called dopants, are introduced deliberately in making a MOSFET, to control the MOSFET electrical behavior.

The *electron mobility* and *hole mobility* columns show the average speed that electrons and holes diffuse through the semiconductor material with an electric field of 1 volt per meter applied across the material. In general, the higher the electron mobility the faster the transistor can operate. The table indicates that Ge is a better material than Si in this respect. However, Ge has four major shortcomings compared to silicon and gallium arsenide:

- Its maximum temperature is limited;

- it has relatively high leakage current;

- it cannot withstand high voltages;

- it is less suitable for fabricating integrated circuits.

Because the electron mobility is higher than the hole mobility for all semiconductor materials, a given bipolar n–p–n transistor tends to be swifter than an equivalent p–n–p transistor. GaAs has the highest electron mobility of the three semiconductors. It is for this reason that GaAs is used in high-frequency applications. A relatively recent FET development, the *high-electron-mobility transistor* (HEMT), has a heterostructure (junction between different semiconductor materials) of aluminium gallium arsenide (AlGaAs)-gallium arsenide (GaAs) which has twice the electron mobility of a GaAs-metal barrier junction. Because of their high speed and low noise, HEMTs are used in satellite receivers working at frequencies around 12 GHz. HEMTs based on gallium nitride and aluminium gallium nitride (AlGaN/GaN HEMTs) provide a still higher electron mobility and are being developed for various applications.

Max. junction temperature values represent a cross section taken from various manufacturers' data sheets. This temperature should not be exceeded or the transistor may be damaged.

Al–Si junction refers to the high-speed (aluminum–silicon) metal–semiconductor barrier diode, commonly known as a Schottky diode. This is included in the table because some silicon power IGFETs have a *parasitic* reverse Schottky diode formed between the source and drain as part of the fabrication process. This diode can be a nuisance, but sometimes it is used in the circuit.

Packaging

Assorted discrete transistors.

Discrete transistors are individually packaged transistors. Transistors come in many different semiconductor packages. The two main categories are *through-hole* (or *leaded*), and *surface-mount*, also known as *surface-mount device* (SMD). The *ball grid array* (BGA) is the latest surface-mount package (currently only for large integrated circuits). It has solder "balls" on the underside in place of leads. Because they are smaller and have shorter interconnections, SMDs have better high-frequency characteristics but lower power rating.

Transistor packages are made of glass, metal, ceramic, or plastic. The package often dictates the power rating and frequency characteristics. Power transistors have larger packages that can be clamped to heat sinks for enhanced cooling. Additionally, most power transistors have the collector or drain physically connected to the metal enclosure. At the other extreme, some surface-mount *microwave* transistors are as small as grains of sand.

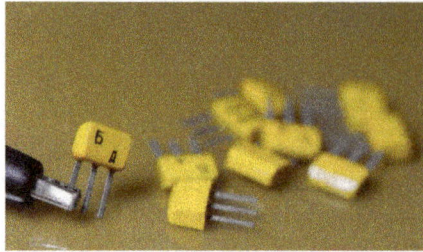

Soviet KT315b transistors.

Often a given transistor type is available in several packages. Transistor packages are mainly standardized, but the assignment of a transistor's functions to the terminals is not: other transistor types can assign other functions to the package's terminals. Even for the same transistor type the terminal assignment can vary (normally indicated by a suffix letter to the part number, q.e. BC212L and BC212K).

Nowadays most transistors come in a wide range of SMT packages, in comparison the list of available through-hole packages is relatively small, here is a short list of the most common through-hole transistors packages in alphabetical order: ATV, E-line, MRT, HRT, SC-43, SC-72, TO-3, TO-18, TO-39, TO-92, TO-126, TO220, TO247, TO251, TO262, ZTX851

Flexible Transistors

Researchers have made several kinds of flexible transistors, including organic field-effect transistors. Flexible transistors are useful in some kinds of flexible displays and other flexible electronics.

References

- Arendt Wintrich, Ulrich Nicolai, Werner Tursky, Tobias Reimann (2010) (in German), [PDF-Version Applikationshandbuch 2015] (2. ed.), ISLE Verlag, ISBN 978-3-938843-83-3.

- Arendt Wintrich; Ulrich Nicolai; Werner Tursky; Tobias Reimann (2015). [PDF-Version Application Manual 2015] Check |url= value (help) (PDF) (2. ed.). ISLE Verlag. ISBN 978-3-938843-83-3.

- Crecraft, David; Stephen Gergely (2002). Analog Electronics: Circuits, Systems and Signal Processing. Butterworth-Heinemann. p. 110. ISBN 0-7506-5095-8.

- Horowitz, Paul; Winfield Hill (1989). The Art of Electronics, 2nd Ed. London: Cambridge University Press. p. 44. ISBN 0-521-37095-7.

- Diffenderfer, Robert (2005). Electronic Devices: Systems and Applications. Thomas Delmar Learning. pp. 95–100. ISBN 1401835147. Retrieved July 22, 2014.

- Horowitz, Paul; Hill, Winfield (1989). The Art of Electronics (2nd ed.). Cambridge University Press. pp. 68–69. ISBN 0-521-37095-7.

- Berlin, Leslie (2005). The Man Behind the Microchip: Robert Noyce and the Invention of Silicon Valley. Oxford University Press. ISBN 0-19-516343-5.

- L.W. Turner,(ed), Electronics Engineer's Reference Book, 4th ed. Newnes-Butterworth, London 1976 ISBN 0-40-800168-2 pp. 8-18

- Christiansen, Donald; Alexander, Charles K. (2005); Standard Handbook of Electrical Engineering (5th ed.). McGraw-Hill, ISBN 0-07-138421-9

- Chelikowski, J. (2004) "Introduction: Silicon in all its Forms", p. 1 in Silicon: evolution and future of a technology. P. Siffert and E. F. Krimmel (eds.). Springer, ISBN 3-540-40546-1.

- McFarland, Grant (2006) Microprocessor design: a practical guide from design planning to manufacturing. McGraw-Hill Professional. p. 10. ISBN 0-07-145951-0.

- Heywang, W. and Zaininger, K. H. (2004) "Silicon: The Semiconductor Material", p. 36 in Silicon: evolution and future of a technology. P. Siffert and E. F. Krimmel (eds.). Springer, 2004 ISBN 3-540-40546-1.

- Zhong Yuan Chang, Willy M. C. Sansen, Low-Noise Wide-Band Amplifiers in Bipolar and CMOS Technologies, page 31, Springer, 1991 ISBN 0792390962.

- Sedra, A.S. & Smith, K.C. (2004). Microelectronic circuits (Fifth ed.). New York: Oxford University Press. p. 397 and Figure 5.17. ISBN 0-19-514251-9.

- Kaplan, Daniel (2003). Hands-On Electronics. New York: Cambridge University Press. pp. 47–54, 60–61. ISBN 978-0-511-07668-8.

- "Milestones:Invention of the First Transistor at Bell Telephone Laboratories, Inc., 1947". IEEE Global History Network. IEEE. Retrieved December 7, 2014.

- Lowe, Doug (2013). "Electronics Components: Diodes". Electronics All-In-One Desk Reference For Dummies. John Wiley & Sons. Retrieved January 4, 2013.

4

Converters Used in Power Electronics

Inverters are appliances that convert power from DC to AC. This chapter studies the power inverter, AC/AC converter (its types like cycloconverter and sparse matrix converter, solar inverter, DC-to-DC converter and rectifier. The content explores the circuit description, size, applications and input and output of each. Power electronics is best understood in confluence with the major topics listed in the following chapter.

Power Inverter

An inverter on a free-standing solar plant

The input voltage, output voltage and frequency, and overall power handling depend on the design of the specific device or circuitry. The inverter does not produce any power; the power is provided by the DC source.

A power inverter, or inverter, is an electronic device or circuitry that changes direct current (DC) to alternating current (AC).

A power inverter can be entirely electronic or may be a combination of mechanical effects (such as a rotary apparatus) and electronic circuitry. Static inverters do not use moving parts in the conversion process.

Input and Output

Input Voltage

A typical power inverter device or circuit requires a relatively stable DC power source capable of supplying enough current for the intended power demands of the system. The input voltage depends on the design and purpose of the inverter. Examples include:

- 12 VDC, for smaller consumer and commercial inverters that typically run from a rechargeable 12 V lead acid battery.

- 24 and 48 VDC, which are common standards for home energy systems.

- 200 to 400 VDC, when power is from photovoltaic solar panels.

- 300 to 450 VDC, when power is from electric vehicle battery packs in vehicle-to-grid systems.

- Hundreds of thousands of volts, where the inverter is part of a high voltage direct current power transmission system.

Output Waveform

An inverter can produce a square wave, modified sine wave, pulsed sine wave, pulse width modulated wave (PWM) or sine wave depending on circuit design. The two dominant commercialized waveform types of inverters as of 2007 are modified sine wave and sine wave.

There are two basic designs for producing household plug-in voltage from a lower-voltage DC source, the first of which uses a switching boost converter to produce a higher-voltage DC and then converts to AC. The second method converts DC to AC at battery level and uses a line-frequency transformer to create the output voltage.

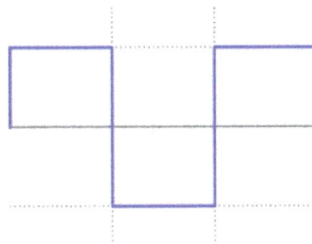

Square wave

Square Wave

This is one of the simplest waveforms an inverter design can produce and is best suited to low-sensitivity applications such as lighting and heating. Square wave output can produce "humming" when connected to audio equipment and is generally unsuitable for sensitive electronics.

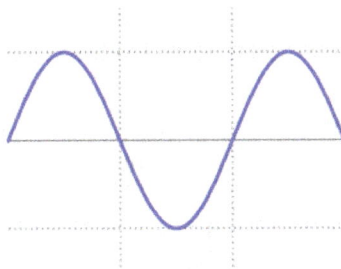

Sine wave

Sine Wave

A power inverter device which produces a multiple step sinusoidal AC waveform is referred to as a *sine wave inverter*. To more clearly distinguish the inverters with outputs of much less distortion than the "modified sine wave" (three step) inverter designs, the manufacturers often use the phrase *pure sine wave inverter*. Almost all consumer grade inverters that are sold as a "pure sine wave inverter" do not produce a smooth sine wave output at all, just a less choppy output than the square wave (one step) and modified sine wave (three step) inverters. In this sense, the phrases "Pure sine wave" or "sine wave inverter" are misleading to the consumer. However, this is not critical for most electronics as they deal with the output quite well.

Where power inverter devices substitute for standard line power, a sine wave output is desirable because many electrical products are engineered to work best with a sine wave AC power source. The standard electric utility power is a sine wave.

Sine wave inverters with more than three steps in the wave output are more complex and have significantly higher cost than a modified sine wave, with only three steps, or square wave (one step) types of the same power handling. Switch-mode power supply (SMPS) devices, such as personal computers or DVD players, function on quality modified sine wave power. AC motors directly operated on non-sinusoidal power may produce extra heat, may have different speed-torque characteristics, or may produce more audible noise than when running on sinusoidal power.

Modified Sine Wave

A *modified sine wave* inverter has a non-square waveform that is a useful approximation of a sine wave for power translation purposes.

Most inexpensive consumer power inverters produce a modified sine wave rather than a pure sine wave.

The waveform in commercially available modified-sine-wave inverters is a square wave with a pause before the polarity reversal, which only needs to cycle back and forth through a three-position switch that outputs forward, off, and reverse output at the pre-determined frequency. Switching states are developed for positive, negative and zero voltages as per the patterns given in the switching Table 2. The peak voltage to RMS voltage ratio does not maintain the same relationship as for a sine wave. The DC bus voltage may be actively regulated, or the "on" and "off" times can be modified to maintain the same RMS value output up to the DC bus voltage to compensate for DC bus voltage variations.

The ratio of on to off time can be adjusted to vary the RMS voltage while maintaining a constant frequency with a technique called pulse width modulation (PWM). The generated gate pulses are given to each switch in accordance with the developed pattern to obtain the desired output. Harmonic spectrum in the output depends on the width of the pulses and the modulation frequency. When operating induction motors, voltage harmonics are usually not of concern; however, harmonic distortion in the current waveform introduces additional heating and can produce pulsating torques.

Numerous items of electric equipment will operate quite well on modified sine wave power inverter devices, especially loads that are resistive in nature such as traditional incandescent light bulbs.

However, the load may operate less efficiently owing to the harmonics associated with a modified sine wave and produce a humming noise during operation. This also affects the efficiency of the system as a whole, since the manufacturer's nominal conversion efficiency does not account for harmonics. Therefore, pure sine wave inverters may provide significantly higher efficiency than modified sine wave inverters.

Most AC motors will run on MSW inverters with an efficiency reduction of about 20% owing to the harmonic content. However, they may be quite noisy. A series LC filter tuned to the fundamental frequency may help.

A common modified sine wave inverter topology found in consumer power inverters is as follows:

An onboard microcontroller rapidly switches on and off power MOSFETs at high frequency like ~50 kHz. The MOSFETs directly pull from a low voltage DC source (such as a battery). This signal then goes through step-up transformers (generally many smaller transformers are placed in parallel to reduce the overall size of the inverter) to produce a higher voltage signal. The output of the step-up transformers then gets filtered by capacitors to produce a high voltage DC supply. Finally, this DC supply is pulsed with additional power MOSFETs by the microcontroller to produce the final modified sine wave signal.

Other Waveforms

By definition there is no restriction on the type of AC waveform an inverter might produce that would find use in a specific or special application.

Output Frequency

The AC output frequency of a power inverter device is usually the same as standard power line frequency, 50 or 60 hertz

If the output of the device or circuit is to be further conditioned (for example stepped up) then the frequency may be much higher for good transformer efficiency.

Output Voltage

The AC output voltage of a power inverter is often regulated to be the same as the grid line voltage, typically 120 or 240 VAC, even when there are changes in the load that the inverter is driving. This allows the inverter to power numerous devices designed for standard line power.

Some inverters also allow selectable or continuously variable output voltages.

Output Power

A power inverter will often have an overall power rating expressed in watts or kilowatts. This describes the power that will be available to the device the inverter is driving and, indirectly, the power that will be needed from the DC source. Smaller popular consumer and commercial devices designed to mimic line power typically range from 150 to 3000 watts.

Not all inverter applications are solely or primarily concerned with power delivery; in some cases the frequency and or waveform properties are used by the follow-on circuit or device.

Batteries

The runtime of an inverter is dependent on the battery power and the amount of power being drawn from the inverter at a given time. As the amount of equipment using the inverter increases, the runtime will decrease. In order to prolong the runtime of an inverter, additional batteries can be added to the inverter.

When attempting to add more batteries to an inverter, there are two basic options for installation:

Series configuration

> If the goal is to increase the overall voltage of the inverter, one can daisy chain batteries in a series configuration. In a series configuration, if a single battery dies, the other batteries will not be able to power the load.

Parallel configuration

> If the goal is to increase capacity and prolong the runtime of the inverter, batteries can be connected in parallel. This increases the overall ampere-hour (Ah) rating of the battery set.

> If a single battery is discharged though, the other batteries will then discharge through it. This can lead to rapid discharge of the entire pack, or even an over-current and possible fire. To avoid this, large paralleled batteries may be connected via diodes or intelligent monitoring with automatic switching to isolate an under-voltage battery from the others.

Applications

DC Power Source Usage

Inverter designed to provide 115 VAC from the 12 VDC source provided in an automobile. The unit shown provides up to 1.2 amperes of alternating current, or enough to power two sixty watt light bulbs.

An inverter converts the DC electricity from sources such as batteries or fuel cells to AC electricity. The electricity can be at any required voltage; in particular it can operate AC equipment designed for mains operation, or rectified to produce DC at any desired voltage.

Uninterruptible Power Supplies

An uninterruptible power supply (UPS) uses batteries and an inverter to supply AC power when mains power is not available. When mains power is restored, a rectifier supplies DC power to recharge the batteries.

Electric Motor Speed Control

Inverter circuits designed to produce a variable output voltage range are often used within motor speed controllers. The DC power for the inverter section can be derived from a normal AC wall outlet or some other source. Control and feedback circuitry is used to adjust the final output of the inverter section which will ultimately determine the speed of the motor operating under its mechanical load. Motor speed control needs are numerous and include things like: industrial motor driven equipment, electric vehicles, rail transport systems, and power tools. Switching states are developed for positive, negative and zero voltages as per the patterns given in the switching Table 1. The generated gate pulses are given to each switch in accordance with the developed pattern and thus the output is obtained.

Power Grid

Grid-tied inverters are designed to feed into the electric power distribution system. They transfer synchronously with the line and have as little harmonic content as possible. They also need a means of detecting the presence of utility power for safety reasons, so as not to continue to dangerously feed power to the grid during a power outage.

Solar

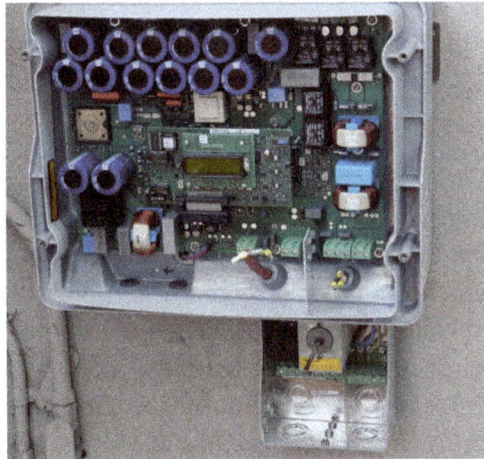

Internal view of a solar inverter. Note the many large capacitors (blue cylinders), used to store energy briefly and improve the output waveform.

A solar inverter is a balance of system (BOS) component of a photovoltaic system and can be used for both, grid-connected and off-grid systems. Solar inverters have special functions adapted for use with photovoltaic arrays, including maximum power point tracking and anti-islanding protection. Solar micro-inverters differ from conventional converters, as an individual micro-converter is attached to each solar panel. This can improve the overall efficiency of the system. The output from several microinverters is then combined and often fed to the electrical grid.

Induction Heating

Inverters convert low frequency main AC power to higher frequency for use in induction heating.

To do this, AC power is first rectified to provide DC power. The inverter then changes the DC power to high frequency AC power. Due to the reduction in the number of DC sources employed, the structure becomes more reliable and the output voltage has higher resolution due to an increase in the number of steps so that the reference sinusoidal voltage can be better achieved. This configuration has recently become very popular in AC power supply and adjustable speed drive applications. This new inverter can avoid extra clamping diodes or voltage balancing capacitors.

There are three kinds of level shifted modulation techniques, namely:

- Phase Opposition Disposition (POD)

- Alternative Phase Opposition Disposition (APOD)

- Phase Disposition (PD)

HVDC Power Transmission

With HVDC power transmission, AC power is rectified and high voltage DC power is transmitted to another location. At the receiving location, an inverter in a static inverter plant converts the power back to AC. The inverter must be synchronized with grid frequency and phase and minimize harmonic generation.

The High voltage DC transmission method can be useful for things like Solar power since solar power is natively DC as it is.

Electroshock Weapons

Electroshock weapons and tasers have a DC/AC inverter to generate several tens of thousands of V AC out of a small 9 V DC battery. First the 9 V DC is converted to 400–2000 V AC with a compact high frequency transformer, which is then rectified and temporarily stored in a high voltage capacitor until a pre-set threshold voltage is reached. When the threshold (set by way of an airgap or TRIAC) is reached, the capacitor dumps its entire load into a pulse transformer which then steps it up to its final output voltage of 20–60 kV. A variant of the principle is also used in electronic flash and bug zappers, though they rely on a capacitor-based voltage multiplier to achieve their high voltage.

Miscellaneous

Typical applications for power inverters include:

- Portable consumer devices that allow the user to connect a battery, or set of batteries, to the device to produce AC power to run various electrical items such as lights, televisions, kitchen appliances, and power tools.

- Use in power generation systems such as electric utility companies or solar generating systems to convert DC power to AC power.

- Use within any larger electronic system where an engineering need exists for deriving an AC source from a DC source.

Circuit Description

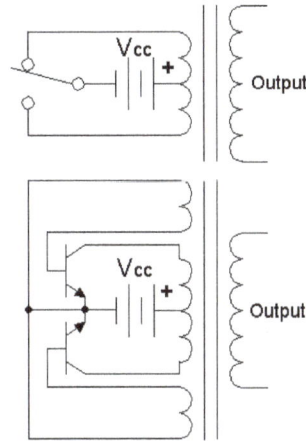

Top: Simple inverter circuit shown with an electromechanical switch *and automatic equivalent* auto-switching device implemented with two transistors and split winding auto-transformer in place of the mechanical switch.

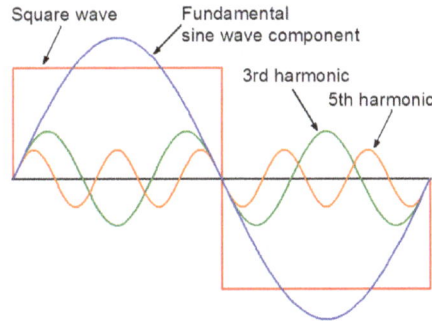

Square waveform with fundamental sine wave component, 3rd harmonic and 5th harmonic

Basic Design

In one simple inverter circuit, DC power is connected to a transformer through the center tap of the primary winding. A switch is rapidly switched back and forth to allow current to flow back to the DC source following two alternate paths through one end of the primary winding and then the other. The alternation of the direction of current in the primary winding of the transformer produces alternating current (AC) in the secondary circuit.

The electromechanical version of the switching device includes two stationary contacts and a spring supported moving contact. The spring holds the movable contact against one of the stationary contacts and an electromagnet pulls the movable contact to the opposite stationary contact. The current in the electromagnet is interrupted by the action of the switch so that the switch continually switches rapidly back and forth. This type of electromechanical inverter switch, called a vibrator or buzzer, was once used in vacuum tube automobile radios. A similar mechanism has been used in door bells, buzzers and tattoo machines.

As they became available with adequate power ratings, transistors and various other types of semiconductor switches have been incorporated into inverter circuit designs. Certain ratings, especially for large systems (many kilowatts) use thyristors (SCR). SCRs provide large power handling capability in a semiconductor device, and can readily be controlled over a variable firing range.

The switch in the simple inverter described above, when not coupled to an output transformer, produces a square voltage waveform due to its simple off and on nature as opposed to the sinusoidal waveform that is the usual waveform of an AC power supply. Using Fourier analysis, periodic waveforms are represented as the sum of an infinite series of sine waves. The sine wave that has the same frequency as the original waveform is called the fundamental component. The other sine waves, called *harmonics*, that are included in the series have frequencies that are integral multiples of the fundamental frequency.

Fourier analysis can be used to calculate the total harmonic distortion (THD). The total harmonic distortion (THD) is the square root of the sum of the squares of the harmonic voltages divided by the fundamental voltage: $\mathrm{THD} = \dfrac{\sqrt{V_2^2 + V_3^2 + V_4^2 + \cdots + V_n^2}}{V_1}$

Advanced Designs

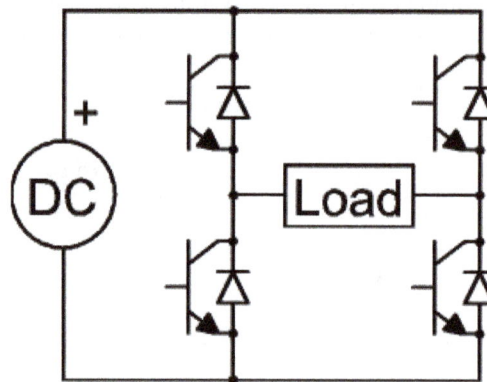

H bridge inverter circuit with transistor switches and antiparallel diodes

There are many different power circuit topologies and control strategies used in inverter designs. Different design approaches address various issues that may be more or less important depending on the way that the inverter is intended to be used.

The issue of waveform quality can be addressed in many ways. Capacitors and inductors can be used to filter the waveform. If the design includes a transformer, filtering can be applied to the primary or the secondary side of the transformer or to both sides. Low-pass filters are applied to allow the fundamental component of the waveform to pass to the output while limiting the passage of the harmonic components. If the inverter is designed to provide power at a fixed frequency, a resonant filter can be used. For an adjustable frequency inverter, the filter must be tuned to a frequency that is above the maximum fundamental frequency.

Since most loads contain inductance, feedback rectifiers or antiparallel diodes are often connected across each semiconductor switch to provide a path for the peak inductive load current when the switch is turned off. The antiparallel diodes are somewhat similar to the *freewheeling diodes* used in AC/DC converter circuits.

Waveform	Signal transitions per period	Harmonics eliminated	Harmonics amplified	System description	THD

	2			2-level square wave	~45%
	4	3, 9, 27, ...		3-level "modified square wave"	>23.8%
	8			5-level "modified square wave"	>6.5%
	10	3, 5, 9, 27	7, 11, ...	2-level very slow PWM	
	12	3, 5, 9, 27	7, 11, ...	3-level very slow PWM	

Fourier analysis reveals that a waveform, like a square wave, that is anti-symmetrical about the 180 degree point contains only odd harmonics, the 3rd, 5th, 7th, etc. Waveforms that have steps of certain widths and heights can attenuate certain lower harmonics at the expense of amplifying higher harmonics. For example, by inserting a zero-voltage step between the positive and negative sections of the square-wave, all of the harmonics that are divisible by three (3rd and 9th, etc.) can be eliminated. That leaves only the 5th, 7th, 11th, 13th etc. The required width of the steps is one third of the period for each of the positive and negative steps and one sixth of the period for each of the zero-voltage steps.

Changing the square wave as described above is an example of pulse-width modulation (PWM). Modulating, or regulating the width of a square-wave pulse is often used as a method of regulating or adjusting an inverter's output voltage. When voltage control is not required, a fixed pulse width can be selected to reduce or eliminate selected harmonics. Harmonic elimination techniques are generally applied to the lowest harmonics because filtering is much more practical at high frequencies, where the filter components can be much smaller and less expensive. *Multiple pulse-width* or *carrier based* PWM control schemes produce waveforms that are composed of many narrow pulses. The frequency represented by the number of narrow pulses per second is called the *switching frequency* or *carrier frequency*. These control schemes are often used in variable-frequency motor control inverters because they allow a wide range of output voltage and frequency adjustment while also improving the quality of the waveform.

Multilevel inverters provide another approach to harmonic cancellation. Multilevel inverters provide an output waveform that exhibits multiple steps at several voltage levels. For example, it is possible to produce a more sinusoidal wave by having split-rail direct current inputs at two voltages, or positive and negative inputs with a central ground. By connecting the inverter output terminals in sequence between the positive rail and ground, the positive rail and the negative rail, the ground rail and the negative rail, then both to the ground rail, a stepped waveform is generated at the inverter output. This is an example of a three level inverter: the two voltages and ground.

More on Achieving a Sine Wave

Resonant inverters produce sine waves with LC circuits to remove the harmonics from a simple square wave. Typically there are several series- and parallel-resonant LC circuits, each tuned to a different harmonic of the power line frequency. This simplifies the electronics, but the inductors and capacitors tend to be large and heavy. Its high efficiency makes this approach popular in large uninterruptible power supplies in data centers that run the inverter continuously in an "online" mode to avoid any switchover transient when power is lost.

A closely related approach uses a ferroresonant transformer, also known as a constant voltage transformer, to remove harmonics and to store enough energy to sustain the load for a few AC cycles. This property makes them useful in standby power supplies to eliminate the switchover transient that otherwise occurs during a power failure while the normally idle inverter starts and the mechanical relays are switching to its output.

Enhanced Quantization

A proposal suggested in *Power Electronics* magazine utilizes two voltages as an improvement over the common commercialized technology which can only apply DC bus voltage in either directions or turn it off. The proposal adds an additional voltage to this design. Each cycle consists of sequence as: v1, v2, v1, 0, −v1, −v2, −v1.

Three-Phase Inverters

Three-phase inverters are used for variable-frequency drive applications and for high power applications such as HVDC power transmission. A basic three-phase inverter consists of three single-phase inverter switches each connected to one of the three load terminals. For the most basic control scheme, the operation of the three switches is coordinated so that one switch operates at each 60 degree point of the fundamental output waveform. This creates a line-to-line output waveform that has six steps. The six-step waveform has a zero-voltage step between the positive and negative sections of the square-wave such that the harmonics that are multiples of three are eliminated as described above. When carrier-based PWM techniques are applied to six-step waveforms, the basic overall shape, or *envelope*, of the waveform is retained so that the 3rd harmonic and its multiples are cancelled.

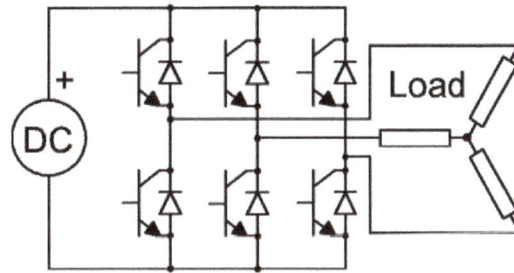

Three-phase inverter with wye connected load

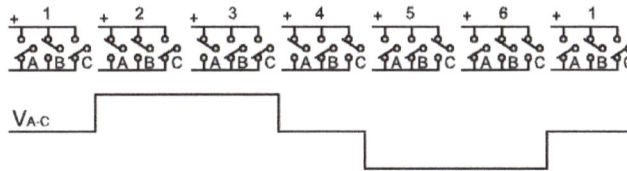

3-phase inverter switching circuit showing 6-step switching sequence and waveform of voltage between terminals A and C ($2^3 - 2$ states)

To construct inverters with higher power ratings, two six-step three-phase inverters can be connected in parallel for a higher current rating or in series for a higher voltage rating. In either case, the output waveforms are phase shifted to obtain a 12-step waveform. If additional inverters are combined, an 18-step inverter is obtained with three inverters etc. Although inverters are usually combined for the purpose of achieving increased voltage or current ratings, the quality of the waveform is improved as well.

Size

Compared to other household electric devices, inverters are large in size and volume. In 2014 Google together with IEEE started an open competition to build a (much) smaller power inverter, with a $1,000,000 prize.

History

Early Inverters

From the late nineteenth century through the middle of the twentieth century, DC-to-AC power conversion was accomplished using rotary converters or motor-generator sets (M-G sets). In the early twentieth century, vacuum tubes and gas filled tubes began to be used as switches in inverter circuits. The most widely used type of tube was the thyratron.

The origins of electromechanical inverters explain the source of the term *inverter*. Early AC-to-DC converters used an induction or synchronous AC motor direct-connected to a generator (dynamo) so that the generator's commutator reversed its connections at exactly the right moments to produce DC. A later development is the synchronous converter, in which the motor and generator windings are combined into one armature, with slip rings at one end and a commutator at the other and only one field frame. The result with either is AC-in, DC-out. With an M-G set, the DC can be considered to be separately generated from the AC; with a synchronous converter, in a certain sense it can be considered to be "mechanically rectified AC". Given the right auxiliary and control equipment, an M-G set or rotary converter can be "run backwards", converting DC to AC. Hence an inverter is an inverted converter.

Controlled Rectifier Inverters

Since early transistors were not available with sufficient voltage and current ratings for most inverter applications, it was the 1957 introduction of the thyristor or silicon-controlled rectifier (SCR) that initiated the transition to solid state inverter circuits.

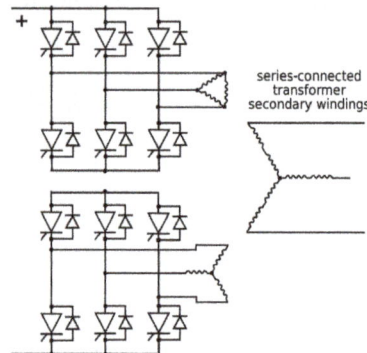

12-pulse line-commutated inverter circuit

The *commutation* requirements of SCRs are a key consideration in SCR circuit designs. SCRs do not turn off or *commutate* automatically when the gate control signal is shut off. They only turn off when the forward current is reduced to below the minimum holding current, which varies with each kind of SCR, through some external process. For SCRs connected to an AC power source, commutation occurs naturally every time the polarity of the source voltage reverses. SCRs connected to a DC power source usually require a means of forced commutation that forces the current to zero when commutation is required. The least complicated SCR circuits employ natural commutation rather than forced commutation. With the addition of forced commutation circuits, SCRs have been used in the types of inverter circuits described above.

In applications where inverters transfer power from a DC power source to an AC power source, it is possible to use AC-to-DC controlled rectifier circuits operating in the inversion mode. In the inversion mode, a controlled rectifier circuit operates as a line commutated inverter. This type of operation can be used in HVDC power transmission systems and in regenerative braking operation of motor control systems.

Another type of SCR inverter circuit is the current source input (CSI) inverter. A CSI inverter is the dual of a six-step voltage source inverter. With a current source inverter, the DC power supply is configured as a current source rather than a voltage source. The inverter SCRs are switched in a six-step sequence to direct the current to a three-phase AC load as a stepped current waveform. CSI inverter commutation methods include load commutation and parallel capacitor commutation. With both methods, the input current regulation assists the commutation. With load commutation, the load is a synchronous motor operated at a leading power factor.

As they have become available in higher voltage and current ratings, semiconductors such as transistors or IGBTs that can be turned off by means of control signals have become the preferred switching components for use in inverter circuits.

Rectifier and Inverter Pulse Numbers

Rectifier circuits are often classified by the number of current pulses that flow to the DC side of the rectifier per cycle of AC input voltage. A single-phase half-wave rectifier is a one-pulse circuit and a single-phase full-wave rectifier is a two-pulse circuit. A three-phase half-wave rectifier is a three-pulse circuit and a three-phase full-wave rectifier is a six-pulse circuit.

With three-phase rectifiers, two or more rectifiers are sometimes connected in series or parallel to obtain higher voltage or current ratings. The rectifier inputs are supplied from special transformers that provide phase shifted outputs. This has the effect of phase multiplication. Six phases are obtained from two transformers, twelve phases from three transformers and so on. The associated rectifier circuits are 12-pulse rectifiers, 18-pulse rectifiers and so on...

When controlled rectifier circuits are operated in the inversion mode, they would be classified by pulse number also. Rectifier circuits that have a higher pulse number have reduced harmonic content in the AC input current and reduced ripple in the DC output voltage. In the inversion mode, circuits that have a higher pulse number have lower harmonic content in the AC output voltage waveform.

Other Notes

The large switching devices for power transmission applications installed until 1970 predominantly used mercury-arc valves. Modern inverters are usually solid state (static inverters). A modern design method features components arranged in an H bridge configuration. This design is also quite popular with smaller-scale consumer devices.

Research

Using 3-D printing and novel semiconductors, researchers at the Department of Energy's Oak Ridge National Laboratory have created a power inverter that could make electric vehicles lighter, more powerful and more efficient.

AC-AC converter

A solid-state AC-AC converter converts an AC waveform to another AC waveform, where the output voltage and frequency can be set arbitrarily.

Categories

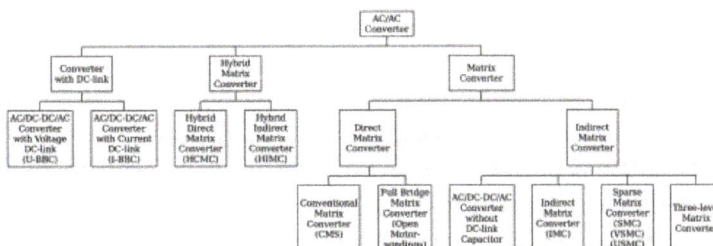

Fig 1: Classification of three-phase AC-AC converter circuits.

Referring to Fig 1, AC-AC converters can be categorized as follows:

Indirect AC-AC (or AC/DC-AC) converters (i.e., with rectifier, DC link and inverter)

- Cycloconverters
- Hybrid matrix converters
- Matrix converters (MC).

DC link converters

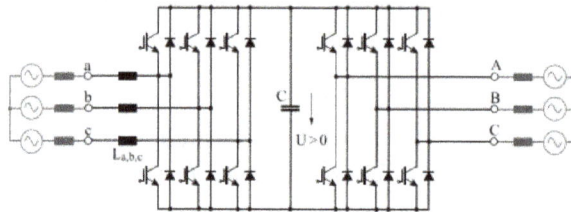

Fig 2: Topology of (regenerative) voltage-source inverter AC/DC-AC converter

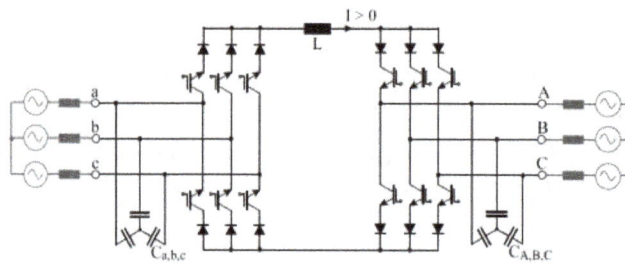

Fig 3: Topology of current-source inverter AC/DC-AC converter

There are two types of converters with DC link:

- Voltage-source inverter (VSI) converters (Fig. 2): In VSI converters, the rectifier consists of a diode-bridge and the DC link consists of a shunt capacitor.

- Current-source inverter (CSI) converters (Fig. 3): In CSI converters, the rectifer consists of a phase-controlled switching device bridge and the DC link consists of 1 or 2 series inductors between one or both legs of the connection between rectifier and inverter.

Any dynamic braking operation required for the motor can be realized by means of braking DC chopper and resistor shunt connected across the rectifier. Alternatively, an anti-parallel thyristor bridge must be provided in the rectifier section to feed energy back into the AC line. Such phase-controlled thyristor-based rectifiers however have higher AC line distortion and lower power factor at low load than diode-based rectifiers.

An AC-AC converter with approximately sinusoidal input currents and bidirectional power flow can be realized by coupling a pulse-width modulation (PWM) rectifier and a PWM inverter to the DC-link. The DC-link quantity is then impressed by an energy storage element that is common to both stages, which is a capacitor C for the voltage DC-link or an inductor L for the current DC-link. The PWM rectifier is controlled in a way that a sinusoidal AC line current is drawn, which is in phase or anti-phase (for energy feedback) with the corresponding AC line phase voltage.

Due to the DC-link storage element, there is the advantage that both converter stages are to a large extent decoupled for control purposes. Furthermore, a constant, AC line independent input quantity exists for the PWM inverter stage, which results in high utilization of the converter's power capability. On the other hand, the DC-link energy storage element has a relatively large physical volume, and when electrolytic capacitors are used, in the case of a voltage DC-link, there is potentially a reduced system lifetime.

Cycloconverters

A cycloconverter constructs an output, variable-frequency, approximately sinusoid waveform by switching segments of the input waveform to the output; there is no intermediate DC link. With switching elements such as SCRs, the output frequency must be lower than the input. Very large cycloconverters (on the order of 10 MW) are manufactured for compressor and wind-tunnel drives, or for variable-speed applications such as cement kilns.

Matrix Converters

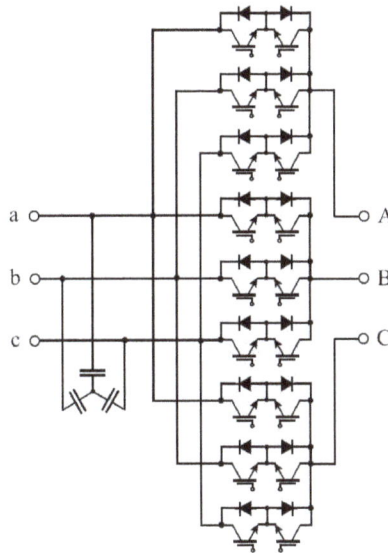

Fig 4: Topology of the Conventional Direct Matrix Converter

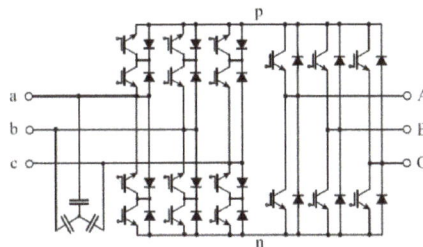

Fig 5: Topology of the indirect matrix converter

In order to achieve higher power density and reliability, it makes sense to consider Matrix Converters that achieve three-phase AC-AC conversion without any intermediate energy storage element. Conventional Direct Matrix Converters (Fig. 4) perform voltage and current conversion in one single stage.

There is the alternative option of indirect energy conversion by employing the Indirect Matrix Converter (Fig. 5) or the Sparse matrix converter which was invented by Prof. Johann W. Kolar from the ETH Zurich. As with the DC-link based VSI and CSI controllers (Fig. 2 and Fig. 3), separate stages are provided for voltage and current conversion, but the DC-link has no intermediate storage element. Generally, by employing matrix converters, the storage element in the DC-link is eliminated at the cost of a larger number of semiconductors. Matrix converters are often seen as a future concept for variable speed drives technology, but despite intensive research over the decades they have until now only achieved low industrial penetration. However, citing recent availability of low-cost, high performance semiconductors, one larger drive manufacturer has over past few years been actively promoting matrix converters.

Various AC/AC Converters

Cycloconverter

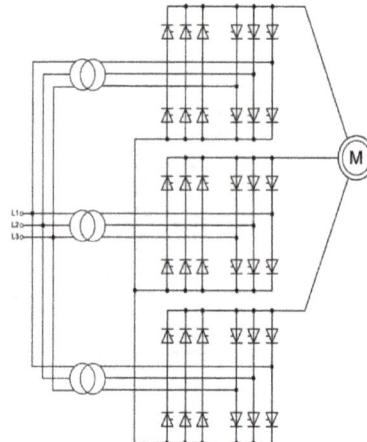

Topology of blocking mode cycloconverter

A cycloconverter (CCV) or a cycloinverter converts a constant voltage, constant frequency AC waveform to another AC waveform of a lower frequency by synthesizing the output waveform from segments of the AC supply without an intermediate DC link (Dorf 1993, pp. 2241–2243 and Lander 1993, p. 181). There are two main types of CCVs, circulating current type or blocking mode type, most commercial high power products being of the blocking mode type.

Characteristics

Whereas phase-controlled SCR switching devices can be used throughout the range of CCVs, low cost, low-power TRIAC-based CCVs are inherently reserved for resistive load applications. The amplitude and frequency of converters' output voltage are both variable. The output to input frequency ratio of a three-phase CCV must be less than about one-third for circulating current mode CCVs or one-half for blocking mode CCVs.(Lander 1993, p. 188) Output waveform quality improves as the *pulse number* of switching-device bridges in phase-shifted configuration increases in CCV's input. In general, CCVs can be with 1-phase/1-phase, 3-phase/1-phase and 3-phase/3-phase input/output configurations, most applications however being 3-phase/3-phase.

Applications

The competitive power rating span of standardized CCVs ranges from few megawatts up to many tens of megawatts. CCVs are used for driving mine hoists, rolling mill main motors, ball mills for ore processing, cement kilns, ship propulsion systems, slip power recovery wound-rotor induction motors (i.e., Scherbius drives) and aircraft 400 Hz power generation. The variable-frequency output of a cycloconverter can be reduced essentially to zero. This means that very large motors can be started on full load at very slow revolutions, and brought gradually up to full speed. This is invaluable with, for example, ball mills, allowing starting with a full load rather than the alternative of having to start the mill with an empty barrel then progressively load it to full capacity. A fully loaded "hard start" for such equipment would essentially be applying full power to a stalled motor. Variable speed and reversing are essential to processes such as hot-rolling steel mills. Previously, SCR-controlled DC motors were used, needing regular brush/commutator servicing and delivering lower efficiency. Cycloconverter-driven synchronous motors need less maintenance and give greater reliability and efficiency. Single-phase bridge CCVs have also been used extensively in electric traction applications to for example produce 25 Hz power in the U.S. and 16 2/3 Hz power in Europe.

Whereas phase-controlled converters including CCVs are gradually being replaced by faster PWM self-controlled converters based on IGBT, GTO, IGCT and other switching devices, these older classical converters are still used at the higher end of the power rating range of these applications.

Harmonics

CCV operation creates current and voltage *harmonics* on the CCV's input and output. AC line harmonics are created on CCV's input accordance to the equation,

- $f_h = f_i (kq\pm1) \pm 6nf_o,$

where

- f_h = harmonic frequency imposed on the AC line
- k and n = integers
- q = pulse number (6, 12 . . .)
- f_o = output frequency of the CCV

Equation's 1st term represents the *pulse number* converter harmonic components starting with six-pulse configuration

Equation's 2nd term denotes the converter's sideband characteristic frequencies including associated interharmonics and subharmonics.

Sparse Matrix Converter

The Sparse Matrix Converter is an AC/AC converter which offers a reduced number of components, a low-complexity modulation scheme, and low realization effort . Invented in 2001 by Prof Johann W. Kolar , sparse matrix converters avoid the multi step commutation procedure of the conventional matrix converter, improving system reliability in industrial operations. Its principal application is in highly compact integrated AC drives.

Characteristics

- Quasi-Direct AC-AC conversion with no DC link energy storage elements

- Sinusoidal input current in phase with mains voltage

- Zero DC link current commutation scheme resulting in lower modulation complexity and very high reliability

- Low complexity of power circuit / power modules available

- Ultra-Sparse Matrix Converter, does show very low realization effort, in case unidirectional power flow can be accepted (admissible displacement of 30° the input current fundamental and input voltage, as well as for the output voltage fundamental and output current), accordingly, a possible application area would be variable speed PSM drives of low dynamics.

Topologies

Matrix Converter

Matrix converter is a device which converts AC input supply to the required variable AC supply as output without any intermediate conversion process whereas in case of Inverter which converts AC - DC - AC which takes more extra components as diode rectifiers, filters, charge-up circuit but not needed those in case of matrix converters

Sparse Matrix Converter

Characteristics of the Sparse Matrix Converter topology are 15 Transistors, 18 Diodes, and 7 Isolated Driver Potentials. Compared to the Direct matrix converter this topology provides identical functionality, but with a reduced number of power switches and the option of employing an improved zero DC-link current commutation scheme, which provides lower control complexity and higher safety and reliability.

Very Sparse Matrix Converter

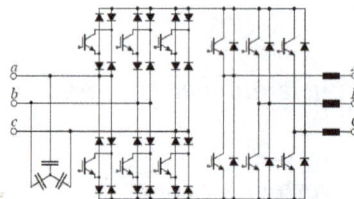

Fig 2: Topology of the very-sparse matrix.

Characteristics of the Very Sparse Matrix Converter topology are 12 Transistors, 30 Diodes, and 10 Isolated Driver Potentials. There are no limitations in functionality compared to the Direct Matrix Converter and Sparse Matrix Converter. Compared to the Sparse Matrix Converter there are fewer transistors but higher conduction losses due to the increased number of diodes in the conduction paths.

Ultra Sparse Matrix Converter

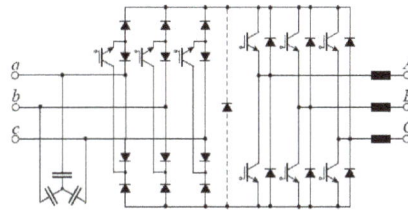

Fig 3: Topology of the ultra-sparse matrix.

Characteristics of the Ultra Sparse Matrix Converter topology are 9 Transistors, 18 Diodes, and 7 Isolated Driver Potentials. The significant limitation of this converter topology compared to the Sparse Matrix Converter is the restriction of its maximal phase displacement between input voltage and input current which is restricted to ± 30°.

Multi-Step Commutation

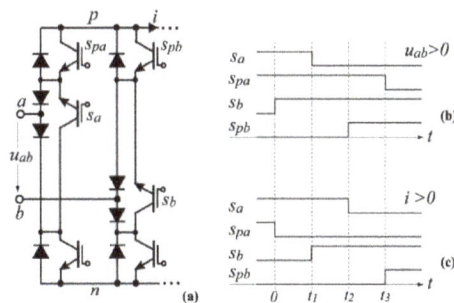

Fig 4: Multistep commutation of the Sparse Matrix Converter rectifier input stage.

This is a commutation scheme, depicted in Fig. 4. For a given switching state of the rectifier input stage, the commutation of the inverter output stage has to be performed in an identical manner to the commutation of a conventional voltage dc-link converter. The basic structure of the commutating bridge legs of the Sparse Matrix Converter is shown in Fig. 4(a). The switch sequence to change the connection of the positive dc-link voltage bus p from input a to input b is shown in Fig. 4(b) and Fig. 4(c). In Fig. 4(b) the assumption is current-independent commutation with uab > 0. In Fig. 4(c) the assumption is voltage-independent commutation with i > 0.

A dead time between the turn-off and turn-on of the power transistors of a bridge leg has to be implemented in order to avoid a short circuit of the dc-link voltage. To change the switching state of the Sparse Matrix Converter rectifier input stage for a given inverter switching state, one has to make sure that there is no bidirectional connection between any two input lines. This guarantees that no short-circuiting of an input line-to-line voltage can occur. Additionally a current path must be continuously provided. Therefore multistep commutation schemes, using voltage independent and current independent commutation as known for the Conventional Direct Matrix Converter , can be employed.

Zero DC Link Current Commutation

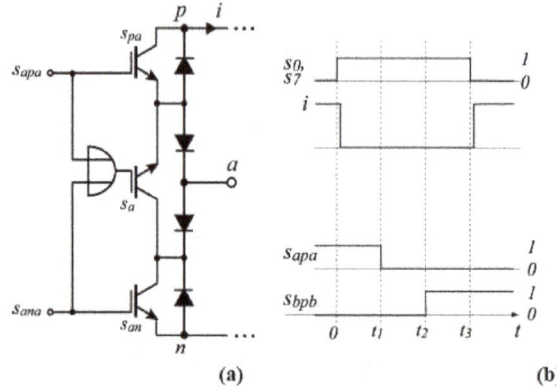

(a)   (b)

Fig 5: Zero DC link current commutation shown for the Sparse Matrix Converter.

The drawback of the multistep commutation describe before is its complexity. Indirect matrix converters like the Sparse Matrix Converter provide a degree of control freedom that is not available for the Conventional Direct Matrix Converter. This can be used to simplify the complex commutation problem. It has been proposed to switch the inverter stage into a free-wheeling state, and then to commutate the rectifier stage with zero dc-link current. This is shown in Fig. 5.

Fig. 5(a) shows the control of the power transistors in one bridge leg of the Sparse Matrix Converter. Fig. 5(b) shows the switching state sequence where s0; s7 = 1 indicates free-wheeling operation of the inverter stage. Furthermore, the dc-link current i is shown.

The zero DC link current commutation scheme gives the additional benefit of a reduction in the switching losses of the input stage. One only has to ensure that no overlapping of turn-on intervals of power transistors in a bridge half occurs, because this would result in a short circuit of an input line-to-line voltage.

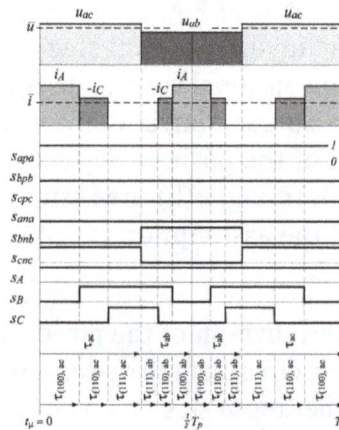

Fig 6: Characteristic voltages and currents and switching states of the Sparse Matrix Converter during on-switching period.

Fig 6 shows the formation of the dc-link voltage u and dc-link current i within one switching period Furthermore, it shows as an example the switching functions of the rectifier and inverter stage for ϕ_1 in interval $(0...\pi/6)$ and ϕ_2 in interval $(0...\pi/6)$. Input stage switching occurs at zero dc-link current. The dc-link current has a constant average value $\overline{i}$ within τ_{ac} and τ_{ab}. The switching state functions are given as s_A, s_B, and s_C. The switching frequency ripple of u_{ac}, u_{ab}, i_A and i_C is neglected

Solar Inverter

Internal view of a solar inverter. Note the many large capacitors (blue cylinders), used to store energy briefly and improve the output waveform.

A solar inverter, or converter or PV inverter, converts the variable direct current (DC) output of a photovoltaic (PV) solar panel into a utility frequency alternating current (AC) that can be fed into a commercial electrical grid or used by a local, off-grid electrical network. It is a critical balance of system (BOS)–component in a photovoltaic system, allowing the use of ordinary AC-powered equipment. Solar power inverters have special functions adapted for use with photovoltaic arrays, including maximum power point tracking and anti-islanding protection.

Classification

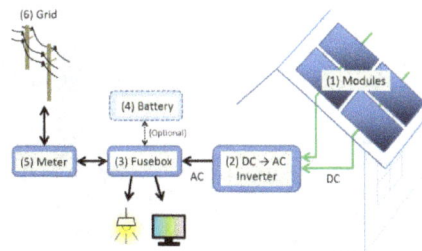

Simplified schematics of a grid-connected residential photovoltaic power system

Solar inverters may be classified into three broad types:

- Stand-alone inverters, used in isolated systems where the inverter draws its DC energy from batteries charged by photovoltaic arrays. Many stand-alone inverters also incorporate integral battery chargers to replenish the battery from an AC source, when available. Normally these do not interface in any way with the utility grid, and as such, are not required to have anti-islanding protection.

- Grid-tie inverters, which match phase with a utility-supplied sine wave. Grid-tie inverters are designed to shut down automatically upon loss of utility supply, for safety reasons. They do not provide backup power during utility outages.

- Battery backup inverters, are special inverters which are designed to draw energy from a battery, manage the battery charge via an onboard charger, and export excess energy to the utility grid. These inverters are capable of supplying AC energy to selected loads during a utility outage, and are required to have anti-islanding protection.

Maximum Power Point Tracking

Solar inverters use *maximum power point tracking* (MPPT) to get the maximum possible power from the PV array. Solar cells have a complex relationship between solar irradiation, temperature and total resistance that produces a non-linear output efficiency known as the *I-V curve*. It is the purpose of the MPPT system to sample the output of the cells and determine a resistance (load) to obtain maximum power for any given environmental conditions.

The fill factor, more commonly known by its abbreviation *FF*, is a parameter which, in conjunction with the open circuit voltage (V_{oc}) and short circuit current (I_{sc}) of the panel, determines the maximum power from a solar cell. Fill factor is defined as the ratio of the maximum power from the solar cell to the product of V_{oc} and I_{sc}.

There are three main types of MPPT algorithms: perturb-and-observe, incremental conductance and constant voltage. The first two methods are often referred to as *hill climbing* methods; they rely on the curve of power plotted against voltage rising to the left of the maximum power point, and falling on the right.

Solar Micro-Inverters

A solar micro-inverter in the process of being installed. The ground wire is attached to the lug and the panel's DC connections are attached to the cables on the lower right. The AC parallel trunk cable runs at the top (just visible).

Solar micro-inverter is an inverter designed to operate with a single PV module. The micro-inverter converts the direct current output from each panel into alternating current. Its design allows parallel connection of multiple, independent units in a modular way.

Micro-inverter advantages include single panel power optimization, independent operation of each panel, plug-and play installation, improved installation and fire safety, minimized costs with system design and stock minimization.

A 2011 study at Appalachian State University reports that individual integrated inverter setup yielded about 20% more power in unshaded conditions and 27% more power in shaded conditions compared to string connected setup using one inverter. Both setups used identical solar panels.

Grid Tied Solar Inverters

Solar grid-tie inverters are designed to quickly disconnect from the grid if the utility grid goes down. This is an NEC requirement that ensures that in the event of a blackout, the grid tie inverter will shut down to prevent the energy it produces from harming any line workers who are sent to fix the power grid.

Grid-tie inverters that are available on the market today use a number of different technologies. The inverters may use the newer high-frequency transformers, conventional low-frequency transformers, or no transformer. Instead of converting direct current directly to 120 or 240 volts AC, high-frequency transformers employ a computerized multi-step process that involves converting the power to high-frequency AC and then back to DC and then to the final AC output voltage.

Historically, there have been concerns about having transformerless electrical systems feed into the public utility grid. The concerns stem from the fact that there is a lack of galvanic isolation between the DC and AC circuits, which could allow the passage of dangerous DC faults to the AC side. Since 2005, the NFPA's NEC allows transformerless (or non-galvanically) inverters. The VDE 0126-1-1 and IEC 6210 also have been amended to allow and define the safety mechanisms needed for such systems. Primarily, residual or ground current detection is used to detect possible fault conditions. Also isolation tests are performed to insure DC to AC separation.

Many solar inverters are designed to be connected to a utility grid, and will not operate when they do not detect the presence of the grid. They contain special circuitry to precisely match the voltage, frequency and phase of the grid.

Solar Charge Controller

A charge controller may be used to power DC equipment with solar panels. The charge controller provides a regulated DC output and stores excess energy in a battery as well as monitoring the battery voltage to prevent under/over charging. More expensive units will also perform maximum power point tracking. An inverter can be connected to the output of a charge controller to drive AC loads.

Solar Pumping Inverters

Advanced solar pumping inverters convert DC voltage from the solar array into AC voltage to drive submersible pumps directly without the need for batteries or other energy storage devices. By utilizing MPPT (maximum power point tracking), solar pumping inverters regulate output frequency to control the speed of the pumps in order to save the pump motor from damage.

Solar pumping inverters usually have multiple ports to allow the input of DC current generated by PV arrays, one port to allow the output of AC voltage, and a further port for input from a water-level sensor.

Market

As of 2014, conversion efficiency for state-of-the-art solar converters reached more than 98 percent. While string inverters are used in residential to medium-sized commercial PV systems, central inverters cover the large commercial and utility-scale market. Market-share for central and string inverters are about 50 percent and 48 percent, respectively, leaving less than 2 percent to micro-inverters.

Inverter/converter market in 2014				
Type	Power	Efficiency[a]	Market share[b]	Remarks
String inverter	up to 100 kW_p [c]	98%	50%	Cost[b] €0.15 per watt-peak. Easy to replace.
Central inverter	above 100 kW_p	98.5%	48%	€0.10 per watt-peak. High reliability. Often sold along with a service contract.
Micro-inverter	module power range	90%–95%	1.5%	€0.40 per watt-peak. Ease of replacement concerns.
DC/DC converter Power optimizer	module power range	98.8%	N/A	€0.40 per watt-peak. Ease of replacement concerns. Inverter is still needed. About 0.75 GW_p installed in 2013.

Source: data by IHS 2014, remarks by Fraunhofer ISE 2014, from: Photovoltaics Report, updated as per 8 September 2014, p. 35, PDF
Notes: [a]best efficiencies displayed, [b]market-share and cost per watt are estimated, [c]kW_p = kilowatt-peak

DC-to-DC converter

A DC-to-DC converter is an electronic circuit or electromechanical device that converts a source of direct current (DC) from one voltage level to another. It is a type of electric power converter. Power levels range from very low (small batteries) to very high (high-voltage power transmission).

History

Before the development of power semiconductors and allied technologies, one way to convert the voltage of a DC supply to a higher voltage, for low-power applications, was to convert it to AC by using a vibrator, followed by a step-up transformer and rectifier. For higher power an electric motor was used to drive a generator of the desired voltage (sometimes combined into a single "dynamotor" unit, a motor and generator combined into one unit, with one winding driving the motor and the other generating the output voltage). These were relatively inefficient and expensive procedures used only when there was no alternative, as to power a car radio (which then used thermionic valves/tubes requiring much higher voltages than available from a 6 or 12 V car battery). The introduction of power semiconductors and integrated circuits made it economically viable to use techniques as described below, for example to convert the DC power supply to high-frequency AC, use a transformer—small, light, and cheap due to the high frequency—to change the voltage, and rectify back to DC. Although by 1976 transistor car radio receivers did not require high voltages, some amateur radio operators continued to use vibrator supplies and dynamotors for mobile transceivers requiring high voltages, although transistorised power supplies were available.

While it was possible to derive a *lower* voltage from a higher with a linear electronic circuit, or even a resistor, these methods dissipated the excess as heat; energy-efficient conversion only became possible with solid-state switch-mode circuits.

Uses

DC to DC converters are used in portable electronic devices such as cellular phones and laptop computers, which are supplied with power from batteries primarily. Such electronic devices often contain several sub-circuits, each with its own voltage level requirement different from that supplied by the battery or an external supply (sometimes higher or lower than the supply voltage). Additionally, the battery voltage declines as its stored energy is drained. Switched DC to DC converters offer a method to increase voltage from a partially lowered battery voltage thereby saving space instead of using multiple batteries to accomplish the same thing.

Most DC to DC converter circuits also regulate the output voltage. Some exceptions include high-efficiency LED power sources, which are a kind of DC to DC converter that regulates the current through the LEDs, and simple charge pumps which double or triple the output voltage.

DC to DC converters developed to maximize the energy harvest for photovoltaic systems and for wind turbines are called power optimizers.

Transformers used for voltage conversion at mains frequencies of 50–60 Hz must be large and heavy for powers exceeding a few watts. This makes them expensive, and they are subject to energy losses in their windings and due to eddy currents in their cores. DC-to-DC techniques that use transformers or inductors work at much higher frequencies, requiring only much smaller, lighter, and cheaper wound components. Consequently these techniques are used even where a mains transformer could be used; for example, for domestic electronic appliances it is preferable to rectify mains voltage to DC, use switch-mode techniques to convert it to high-frequency AC at the desired voltage, then, usually, rectify to DC. The entire complex circuit is cheaper and more efficient than a simple mains transformer circuit of the same output.

Electronic Conversion

Linear regulators which are used to output a stable DC independent of input voltage and output load from a higher but less stable input by dissipating excess volt-amperes as heat, could be described literally as DC-to-DC converters, but this is not usual usage. (The same could be said of a simple voltage dropper resistor, whether or not stabilised by a following voltage regulator or Zener diode.)

There are also simple capacitive voltage doubler and Dickson multiplier circuits using diodes and capacitors to multiply a DC voltage by an integer value, typically delivering only a small current.

Practical electronic converters use switching techniques. Switched-mode DC-to-DC converters convert one DC voltage level to another, which may be higher or lower, by storing the input energy temporarily and then releasing that energy to the output at a different voltage. The storage may be in either magnetic field storage components (inductors, transformers) or electric field storage components (capacitors). This conversion method can increase or decrease voltage. Switching conversion is more power efficient (often 75% to 98%) than linear voltage regulation, which dissipates unwanted power as heat. Fast semiconductor device rise and fall times are required for efficiency; however, these fast transitions combine with layout parasitic effects to make circuit design challenging. The higher efficiency of a switched-mode converter reduces the heatsinking needed, and increases battery endurance of portable equipment. Efficiency has improved since the late 1980s due to the use of power FETs, which are able to switch more efficiently with lower switching losses at higher frequencies than power bipolar transistors, and use less complex drive circuitry. Another important improvement in DC-DC converters is replacing the flywheel diode by synchronous rectification using a power FET, whose "on resistance" is much lower, reducing switching losses. Before the wide availability of power semiconductors, low-power DC-to-DC synchronous converters consisted of an electro-mechanical vibrator followed by a voltage step-up transformer feeding a vacuum tube or semiconductor rectifier, or synchronous rectifier contacts on the vibrator.

Most DC-to-DC converters are designed to move power in only one direction, from dedicated input to output. However, all switching regulator topologies can be made bidirectional and able to move power in either direction by replacing all diodes with independently controlled active rectification. A bidirectional converter is useful, for example, in applications requiring regenerative braking of vehicles, where power is supplied *to* the wheels while driving, but supplied *by* the wheels when braking.

Switching converters are electronically complex, although this is embodied in integrated circuits, with few components needed. They need careful design of the circuit and physical layout to reduce switching noise (EMI / RFI) to acceptable levels and, like all high-frequency circuits, for stable operation. Cost was higher than linear regulators in voltage-dropping applications, but this dropped with advances in chip design.

DC-to-DC converters are available as integrated circuits (ICs) requiring few additional components. Converters are also available as complete hybrid circuit modules, ready for use within an electronic assembly.

Magnetic

In these DC-to-DC converters, energy is periodically stored within and released from a magnetic field in an inductor or a transformer, typically within a frequency range of 300 kHz to

10 MHz. By adjusting the duty cycle of the charging voltage (that is, the ratio of the on/off times), the amount of power transferred to a load can be more easily controlled, though this control can also be applied to the input current, the output current, or to maintain constant power. Transformer-based converters may provide isolation between input and output. In general, the term *DC-to-DC converter* refers to one of these switching converters. These circuits are the heart of a switched-mode power supply. Many topologies exist. This table shows the most common ones.

	Forward (energy transfers through the magnetic field)	Flyback (energy is stored in the magnetic field)
No transformer (non-isolated)	☐ Step-down (buck) - The output voltage is lower than the input voltage, and of the same polarity.	☐ Non-inverting: The output voltage is the same polarity as the input. ○ Step-up (boost) - The output voltage is higher than the input voltage. ○ SEPIC - The output voltage can be lower or higher than the input. ☐ Inverting: the output voltage is of the opposite polarity as the input. ○ Inverting (buck-boost). ○ Ćuk - Output current is continuous.
	☐ True buck-boost - The output voltage is the same polarity as the input and can be lower or higher.	
	☐ Split-pi (boost-buck) - Allows bidirectional voltage conversion with the output voltage the same polarity as the input and can be lower or higher.	
With transformer (isolatable)	☐ Forward - 1 or 2 transistor drive. ☐ Push-pull (half bridge) - 2 transistors drive. ☐ Full bridge - 4 transistor drive.	☐ Flyback - 1 transistor drive.

In addition, each topology may be:

Hard switched

Transistors switch quickly while exposed to both full voltage and full current

Resonant

An LC circuit shapes the voltage across the transistor and current through it so that the transistor switches when either the voltage or the current is zero

Magnetic DC-to-DC converters may be operated in two modes, according to the current in its main magnetic component (inductor or transformer):

Continuous

The current fluctuates but never goes down to zero

Discontinuous

> The current fluctuates during the cycle, going down to zero at or before the end of each cycle

A converter may be designed to operate in continuous mode at high power, and in discontinuous mode at low power.

The half bridge and flyback topologies are similar in that energy stored in the magnetic core needs to be dissipated so that the core does not saturate. Power transmission in a flyback circuit is limited by the amount of energy that can be stored in the core, while forward circuits are usually limited by the I/V characteristics of the switches.

Although MOSFET switches can tolerate simultaneous full current and voltage (although thermal stress and electromigration can shorten the MTBF), bipolar switches generally can't so require the use of a snubber (or two).

High-current systems often use multiphase converters, also called interleaved converters. Multiphase regulators can have better ripple and better response times than single-phase regulators.

Many laptop and desktop motherboards include interleaved buck regulators, sometimes as a voltage regulator module.

Capacitive

Switched capacitor converters rely on alternately connecting capacitors to the input and output in differing topologies. For example, a switched-capacitor reducing converter might charge two capacitors in series and then discharge them in parallel. This would produce the same output power (less that lost to efficiency of under 100%) at, ideally, half the input voltage and twice the current. Because they operate on discrete quantities of charge, these are also sometimes referred to as charge pump converters. They are typically used in applications requiring relatively small currents, as at higher currents the increased efficiency and smaller size of switch-mode converters makes them a better choice. They are also used at extremely high voltages, as magnetics would break down at such voltages.

Electromechanical Conversion

A motor generator with separate motor and generator.

A motor-generator set, mainly of historical interest, consists of an electric motor and generator coupled together. A *dynamotor* combines both functions into a single unit with coils for both the motor and the generator functions wound around a single rotor; both coils share the same outer field coils or magnets. Typically the motor coils are driven from a commutator on one end of the shaft, when the generator coils output to another commutator on the other end of the shaft. The entire rotor and shaft assembly is smaller in size than a pair of machines, and may not have any exposed drive shafts.

Motor-generators can convert between any combination of DC and AC voltage and phase standards. Large motor-generator sets were widely used to convert industrial amounts of power while smaller units were used to convert battery power (6, 12 or 24 V DC) to a high DC voltage, which was required to operate vacuum tube (thermionic valve) equipment.

For lower-power requirements at voltages higher than supplied by a vehicle battery, vibrator or "buzzer" power supplies were used. The vibrator oscillated mechanically, with contacts that switched the polarity of the battery many times per second, effectively converting DC to square wave AC, which could then be fed to a transformer of the required output voltage(s). It made a characteristic buzzing noise.

Electrochemical Conversion

A further means of DC to DC conversion in the kilowatts to megawatts range is presented by using redox flow batteries such as the vanadium redox battery.

Chaotic Behavior

DC-to-DC converters are subject to different types of chaotic dynamics such as bifurcation, crisis, and intermittency.

Terminology

Step-down

> A converter where output voltage is lower than the input voltage (such as a buck converter).

Step-up

> A converter that outputs a voltage higher than the input voltage (such as a boost converter).

Continuous current mode

> Current and thus the magnetic field in the inductive energy storage never reach zero.

Discontinuous current mode

> Current and thus the magnetic field in the inductive energy storage may reach or cross zero.

Noise

> Unwanted electrical and electromagnetic signal noise, typically switching artifacts.

RF noise

> Switching converters inherently emit radio waves at the switching frequency and its har-

monics. Switching converters that produce triangular switching current, such as the Split-Pi, forward converter, or Ćuk converter in continuous current mode, produce less harmonic noise than other switching converters. RF noise causes electromagnetic interference (EMI). Acceptable levels depend upon requirements, e.g. proximity to RF circuitry needs more suppression than simply meeting regulations.

Input noise

The input voltage may have non-negligible noise. Additionally, if the converter loads the input with sharp load edges, the converter can emit RF noise from the supplying power lines. This should be prevented with proper filtering in the input stage of the converter.

Output noise

The output of an ideal DC-to-DC converter is a flat, constant output voltage. However, real converters produce a DC output upon which is superimposed some level of electrical noise. Switching converters produce switching noise at the switching frequency and its harmonics. Additionally, all electronic circuits have some thermal noise. Some sensitive radio-frequency and analog circuits require a power supply with so little noise that it can only be provided by a linear regulator. Some analog circuits which require a power supply with relatively low noise can tolerate some of the less-noisy switching converters, e.g. using continuous triangular waveforms rather than square waves.

Rectifier

A rectifier diode (silicon controlled rectifier) and associated mounting hardware. The heavy threaded stud attaches the device to a heatsink to dissipate heat.

A rectifier is an electrical device that converts alternating current (AC), which periodically reverses direction, to direct current (DC), which flows in only one direction. The process is known as rectification. Physically, rectifiers take a number of forms, including vacuum tube diodes, mercury-arc valves, copper and selenium oxide rectifiers, semiconductor diodes, silicon-controlled rectifiers and other silicon-based semiconductor switches. Historically, even synchronous electromechanical switches and motors have been used. Early radio receivers, called crystal radios, used a "cat's whisker" of fine wire pressing on a crystal of galena (lead sulfide) to serve as a point-contact rectifier or "crystal detector".

Rectifiers have many uses, but are often found serving as components of DC power supplies and high-voltage direct current power transmission systems. Rectification may serve in roles other than to generate direct current for use as a source of power. As noted, detectors of radio signals serve as rectifiers. In gas heating systems flame rectification is used to detect presence of a flame.

Because of the alternating nature of the input AC sine wave, the process of rectification alone produces a DC current that, though unidirectional, consists of pulses of current. Many applications of rectifiers, such as power supplies for radio, television and computer equipment, require a *steady* constant DC current (as would be produced by a battery). In these applications the output of the rectifier is smoothed by an electronic filter (usually a capacitor) to produce a steady current.

More complex circuitry that performs the opposite function, converting DC to AC, is called an inverter.

Rectifier Devices

Before the development of silicon semiconductor rectifiers, vacuum tube thermionic diodes and copper oxide- or selenium-based metal rectifier stacks were used. With the introduction of semiconductor electronics, vacuum tube rectifiers became obsolete, except for some enthusiasts of vacuum tube audio equipment. For power rectification from very low to very high current, semiconductor diodes of various types (junction diodes, Schottky diodes, etc.) are widely used.

Other devices that have control electrodes as well as acting as unidirectional current valves are used where more than simple rectification is required—e.g., where variable output voltage is needed. High-power rectifiers, such as those used in high-voltage direct current power transmission, employ silicon semiconductor devices of various types. These are thyristors or other controlled switching solid-state switches, which effectively function as diodes to pass current in only one direction.

Rectifier Circuits

Rectifier circuits may be single-phase or multi-phase (three being the most common number of phases). Most low power rectifiers for domestic equipment are single-phase, but three-phase rectification is very important for industrial applications and for the transmission of energy as DC (HVDC).

Single-Phase Rectifiers

Half-Wave Rectification

In half-wave rectification of a single-phase supply, either the positive or negative half of the AC wave is passed, while the other half is blocked. Because only one half of the input waveform reaches the output, mean voltage is lower. Half-wave rectification requires a single diode in a single-phase supply, or three in a three-phase supply. Rectifiers yield a unidirectional but pulsating direct current; half-wave rectifiers produce far more ripple than full-wave rectifiers, and much more filtering is needed to eliminate harmonics of the AC frequency from the output.

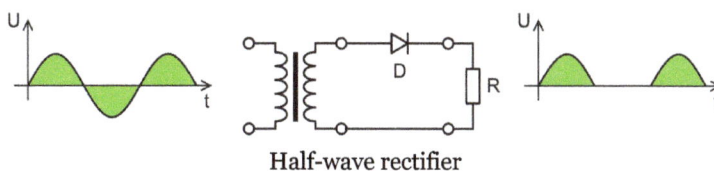

Half-wave rectifier

The no-load output DC voltage of an ideal half-wave rectifier for a sinusoidal input voltage is:

$$V_{rms} = \frac{V_{peak}}{2}$$

$$V_{dc} = \frac{V_{peak}}{\pi}$$

where:

V_{dc}, V_{av} – the DC or average output voltage,

V_{peak}, the peak value of the phase input voltages,

V_{rms}, the root-mean-square value of output voltage.

Full-Wave Rectification

A full-wave rectifier converts the whole of the input waveform to one of constant polarity (positive or negative) at its output. Full-wave rectification converts both polarities of the input waveform to pulsating DC (direct current), and yields a higher average output voltage. Two diodes and a center tapped transformer, or four diodes in a bridge configuration and any AC source (including a transformer without center tap), are needed. Single semiconductor diodes, double diodes with common cathode or common anode, and four-diode bridges, are manufactured as single components.

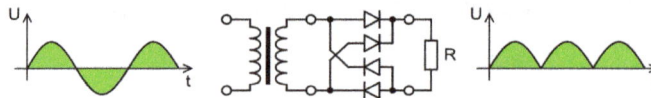

Graetz bridge rectifier: a full-wave rectifier using four diodes.

For single-phase AC, if the transformer is center-tapped, then two diodes back-to-back (cathode-to-cathode or anode-to-anode, depending upon output polarity required) can form a full-wave rectifier. Twice as many turns are required on the transformer secondary to obtain the same output voltage than for a bridge rectifier, but the power rating is unchanged.

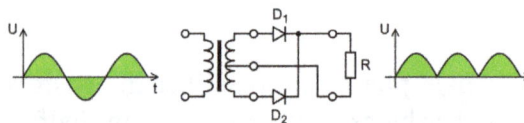

Full-wave rectifier using a center tap transformer and 2 diodes.

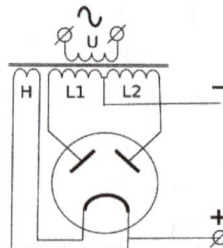

Full-wave rectifier, with vacuum tube having two anodes.

The average and root-mean-square no-load output voltages of an ideal single-phase full-wave rectifier are:

$$V_{dc} = V_{av} = \frac{2V_{peak}}{\pi}$$

$$V_{rms} = \frac{V_{peak}}{\sqrt{2}}$$

Very common double-diode rectifier vacuum tubes contained a single common cathode and two anodes inside a single envelope, achieving full-wave rectification with positive output. The 5U4 and 5Y3 were popular examples of this configuration.

Three-Phase Rectifiers

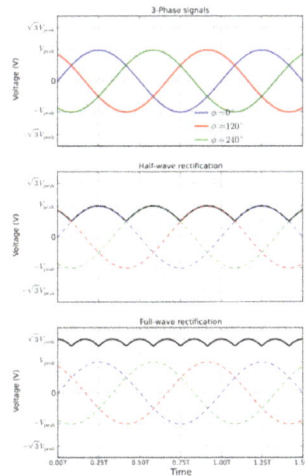

3-phase AC input, half- and full-wave rectified DC output waveforms

Single-phase rectifiers are commonly used for power supplies for domestic equipment. However, for most industrial and high-power applications, three-phase rectifier circuits are the norm. As with single-phase rectifiers, three-phase rectifiers can take the form of a half-wave circuit, a full-wave circuit using a center-tapped transformer, or a full-wave bridge circuit.

Thyristors are commonly used in place of diodes to create a circuit that can regulate the output voltage. Many devices that provide direct current actually *generate* three-phase AC. For example, an automobile alternator contains six diodes, which function as a full-wave rectifier for battery charging.

Three-Phase, Half-Wave Circuit

An uncontrolled three-phase, half-wave circuit requires three diodes, one connected to each phase. This is the simplest type of three-phase rectifier but suffers from relatively high harmonic distortion on both the AC and DC connections. This type of rectifier is said to have a pulse-number of three, since the output voltage on the DC side contains three distinct pulses per cycle of the grid frequency.

Three-Phase, Full-Wave Circuit Using Center-Tapped Transformer

If the AC supply is fed via a transformer with a center tap, a rectifier circuit with improved harmonic performance can be obtained. This rectifier now requires six diodes, one connected to each end of each transformer secondary winding. This circuit has a pulse-number of six, and in effect, can be thought of as a six-phase, half-wave circuit.

Before solid state devices became available, the half-wave circuit, and the full-wave circuit using a center-tapped transformer, were very commonly used in industrial rectifiers using mercury-arc valves. This was because the three or six AC supply inputs could be fed to a corresponding number of anode electrodes on a single tank, sharing a common cathode.

With the advent of diodes and thyristors, these circuits have become less popular and the three-phase bridge circuit has become the most common circuit.

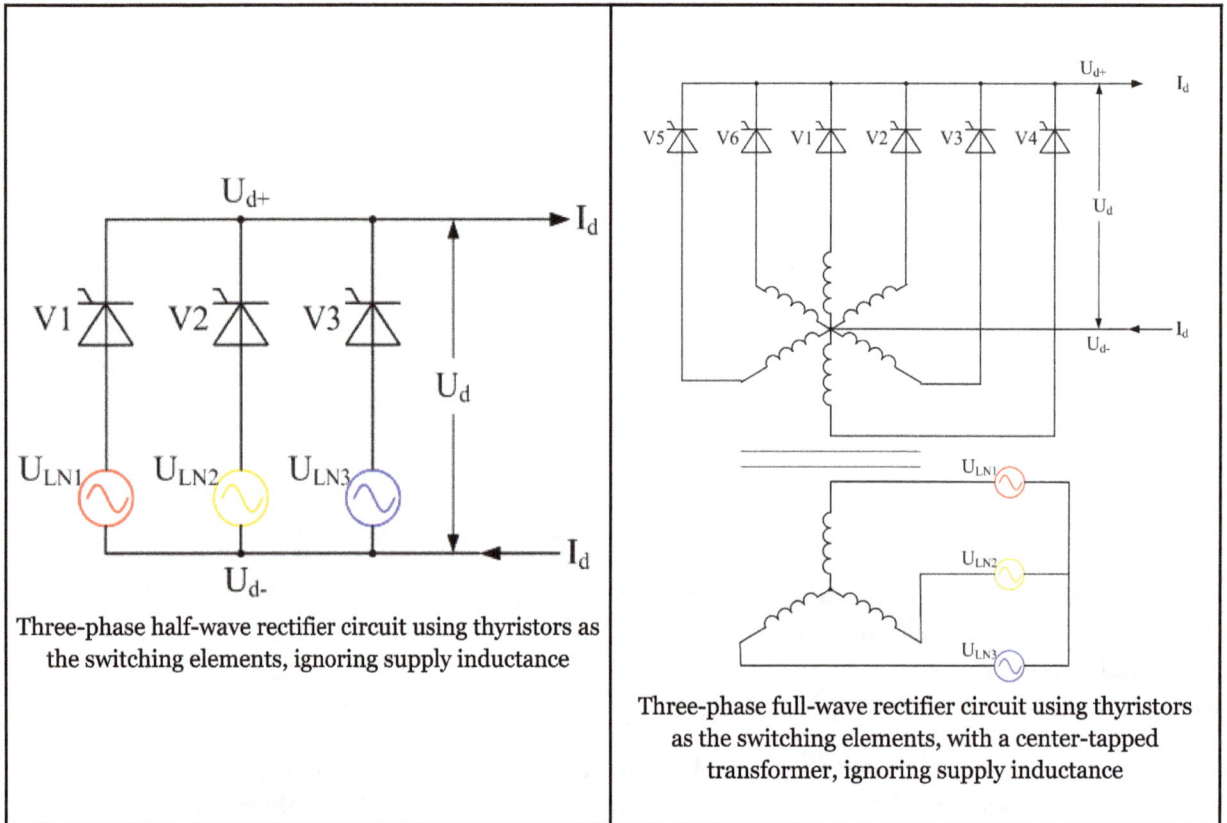

Three-phase half-wave rectifier circuit using thyristors as the switching elements, ignoring supply inductance

Three-phase full-wave rectifier circuit using thyristors as the switching elements, with a center-tapped transformer, ignoring supply inductance

Three-Phase Bridge Rectifier

For an uncontrolled three-phase bridge rectifier, six diodes are used, and the circuit again has a pulse number of six. For this reason, it is also commonly referred to as a six-pulse bridge.

For low-power applications, double diodes in series, with the anode of the first diode connected to the cathode of the second, are manufactured as a single component for this purpose. Some commercially available double diodes have all four terminals available so the user can configure them for single-phase split supply use, half a bridge, or three-phase rectifier.

For higher-power applications, a single discrete device is usually used for each of the six arms of the bridge. For the very highest powers, each arm of the bridge may consist of tens or hundreds of separate devices in parallel (where very high current is needed, for example in aluminium smelting) or in series (where very high voltages are needed, for example in high-voltage direct current power transmission).

Disassembled automobile alternator, showing the six diodes that comprise a full-wave three-phase bridge rectifier.

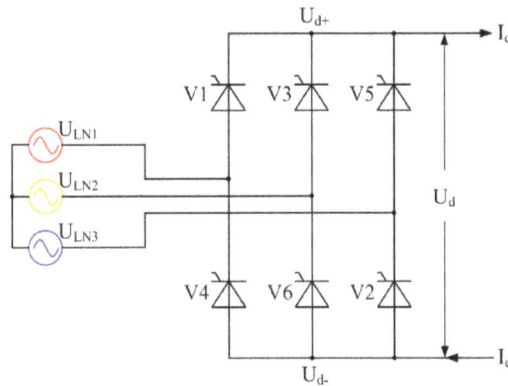

Three-phase full-wave bridge rectifier circuit using thyristors as the switching elements, ignoring supply inductance

For a three-phase full-wave diode rectifier, the ideal, no-load average output voltage is

$$V_{dc} = V_{av} = \frac{3\sqrt{3}V_{peak}}{\pi}$$

If thyristors are used in place of diodes, the output voltage is reduced by a factor $\cos(\alpha)$:

$$V_{dc} = V_{av} = \frac{3\sqrt{3}V_{peak}}{\pi}\cos\alpha$$

Or, expressed in terms of the line to line input voltage:

$$V_{dc} = V_{av} = \frac{3V_{LLpeak}}{\pi}\cos\alpha$$

Introduction to Modern Power Electronics

Where:

V_{LLpeak}, the peak value of the line to line input voltages,

V_{peak}, the peak value of the phase (line to neutral) input voltages,

α, firing angle of the thyristor (0 if diodes are used to perform rectification)

The above equations are only valid when no current is drawn from the AC supply or in the theoretical case when the AC supply connections have no inductance. In practice, the supply inductance causes a reduction of DC output voltage with increasing load, typically in the range 10–20% at full load.

The effect of supply inductance is to slow down the transfer process (called commutation) from one phase to the next. As result of this is that at each transition between a pair of devices, there is a period of overlap during which three (rather than two) devices in the bridge are conducting simultaneously. The overlap angle is usually referred to by the symbol μ (or u), and may be 20 30° at full load.

With supply inductance taken into account, the output voltage of the rectifier is reduced to:

$$V_{dc} = V_{av} = \frac{3V_{\text{LLpeak}}}{\pi} \cos(\alpha + \mu)$$

The overlap angle μ is directly related to the DC current, and the above equation may be re-expressed as:

$$V_{dc} = V_{av} = \frac{3V_{\text{LLpeak}}}{\pi} \cos(\alpha) - 6 f L_c I_d$$

Where:

L_c, the commutating inductance per phase

I_d, the direct current

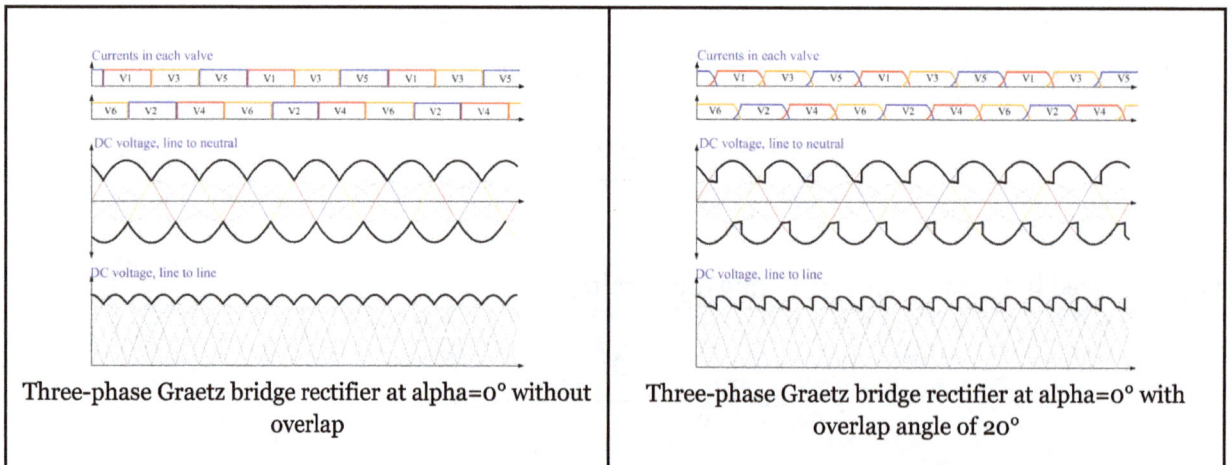

| Three-phase Graetz bridge rectifier at alpha=0° without overlap | Three-phase Graetz bridge rectifier at alpha=0° with overlap angle of 20° |

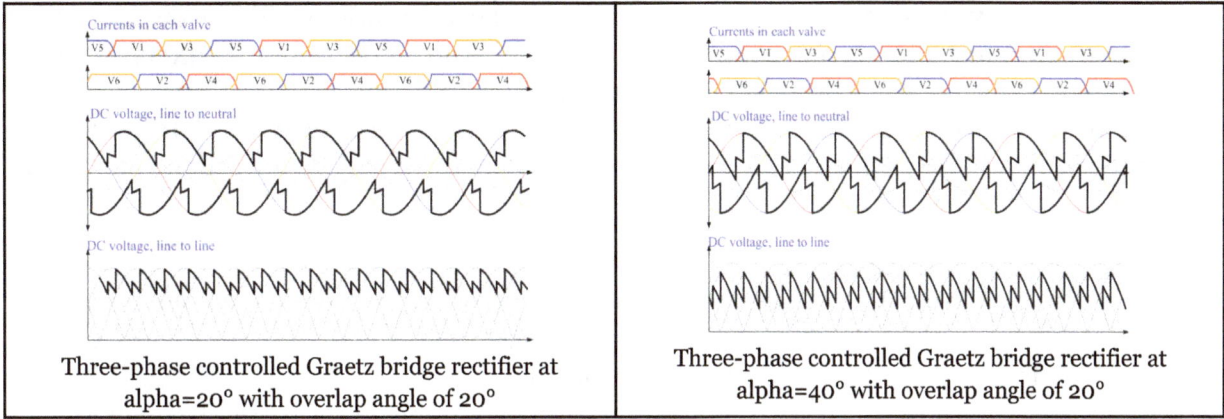

Three-phase controlled Graetz bridge rectifier at
alpha=20° with overlap angle of 20°

Three-phase controlled Graetz bridge rectifier at
alpha=40° with overlap angle of 20°

Twelve-Pulse Bridge

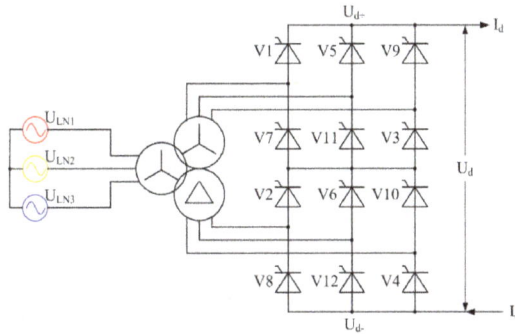

Twelve pulse bridge rectifier using thyristors as the switching elements

Although better than single-phase rectifiers or three-phase half-wave rectifiers, six-pulse rectifier circuits still produce considerable harmonic distortion on both the AC and DC connections.
For very high-power rectifiers the twelve-pulse bridge connection is usually used. A twelve-pulse
bridge consists of two six-pulse bridge circuits connected in series, with their AC connections fed
from a supply transformer that produces a 30° phase shift between the two bridges. This cancels
many of the characteristic harmonics the six-pulse bridges produce.

The 30 degree phase shift is usually achieved by using a transformer with two sets of secondary
windings, one in star (wye) connection and one in delta connection.

Voltage-Multiplying Rectifiers

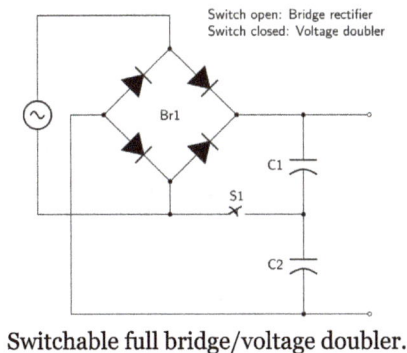

Switchable full bridge/voltage doubler.

The simple half-wave rectifier can be built in two electrical configurations with the diode pointing in opposite directions, one version connects the negative terminal of the output direct to the AC supply and the other connects the positive terminal of the output direct to the AC supply. By combining both of these with separate output smoothing it is possible to get an output voltage of nearly double the peak AC input voltage. This also provides a tap in the middle, which allows use of such a circuit as a split rail power supply.

A variant of this is to use two capacitors in series for the output smoothing on a bridge rectifier then place a switch between the midpoint of those capacitors and one of the AC input terminals. With the switch open, this circuit acts like a normal bridge rectifier. With the switch closed, it act like a voltage doubling rectifier. In other words, this makes it easy to derive a voltage of roughly 320 V (±15%, approx.) DC from any 120 V or 230 V mains supply in the world, this can then be fed into a relatively simple switched-mode power supply. However, for a given desired ripple, the value of both capacitors must be twice the value of the single one required for a normal bridge rectifier; when the switch is closed each one must filter the output of a half-wave rectifier, and when the switch is open the two capacitors are connected in series with an equivalent value of half one of them.

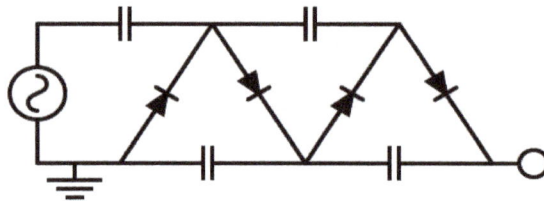

Cockcroft Walton voltage multiplier

Cascaded diode and capacitor stages can be added to make a voltage multiplier (Cockroft-Walton circuit). These circuits are capable of producing a DC output voltage potential tens of times that of the peak AC input voltage, but are limited in current capacity and regulation. Diode voltage multipliers, frequently used as a trailing boost stage or primary high voltage (HV) source, are used in HV laser power supplies, powering devices such as cathode ray tubes (CRT) (like those used in CRT based television, radar and sonar displays), photon amplifying devices found in image intensifying and photo multiplier tubes (PMT), and magnetron based radio frequency (RF) devices used in radar transmitters and microwave ovens. Before the introduction of semiconductor electronics, transformerless powered vacuum tube receivers powered directly from AC power sometimes used voltage doublers to generate about 170 VDC from a 100–120 V power line.

Rectifier Efficiency

Rectifier efficiency (η) is defined as the ratio of DC output power to the input power from the AC supply. Even with ideal rectifiers with no losses, the efficiency is less than 100% because some of the output power is AC power rather than DC which manifests as ripple superimposed on the DC waveform. For a half-wave rectifier efficiency is very poor,

$$P_{in} = \frac{V_{peak}}{2} \cdot \frac{I_{peak}}{2}$$

(the divisors are 2 rather than √2 because no power is delivered on the negative half-cycle)

$$P_{out} = \frac{V_{peak}}{\pi} \cdot \frac{I_{peak}}{\pi}$$

Thus maximum efficiency for a half-wave rectifier is,

$$\eta = \frac{P_{out}}{P_{in}} = \frac{4}{\pi^2} \approx 40.5\%$$

Similarly, for a full-wave rectifier,

$$\eta = \frac{P_{out}}{P_{in}} = \frac{8}{\pi^2} \approx 81.0\%$$

Efficiency is reduced by losses in transformer windings and power dissipation in the rectifier element itself. Efficiency can be improved with the use of smoothing circuits which reduce the ripple and hence reduce the AC content of the output. Three-phase rectifiers, especially three-phase full-wave rectifiers, have much greater efficiencies because the ripple is intrinsically smaller. In some three-phase and multi-phase applications the efficiency is high enough that smoothing circuitry is unnecessary.

Rectifier Losses

A real rectifier characteristically drops part of the input voltage (a voltage drop, for silicon devices, of typically 0.7 volts plus an equivalent resistance, in general non-linear)—and at high frequencies, distorts waveforms in other ways. Unlike an ideal rectifier, it dissipates some power.

An aspect of most rectification is a loss from the peak input voltage to the peak output voltage, caused by the built-in voltage drop across the diodes (around 0.7 V for ordinary silicon p–n junction diodes and 0.3 V for Schottky diodes). Half-wave rectification and full-wave rectification using a center-tapped secondary produces a peak voltage loss of one diode drop. Bridge rectification has a loss of two diode drops. This reduces output voltage, and limits the available output voltage if a very low alternating voltage must be rectified. As the diodes do not conduct below this voltage, the circuit only passes current through for a portion of each half-cycle, causing short segments of zero voltage (where instantaneous input voltage is below one or two diode drops) to appear between each "hump".

Peak loss is very important for low voltage rectifiers (for example, 12 V or less) but is insignificant in high-voltage applications such as HVDC.

Rectifier Output Smoothing

While half-wave and full-wave rectification can deliver unidirectional current, neither produces a constant voltage. Producing steady DC from a rectified AC supply requires a smoothing circuit or filter. In its simplest form this can be just a reservoir capacitor or smoothing capacitor, placed at the DC output of the rectifier. There is still an AC ripple voltage component at the power supply frequency for a half-wave rectifier, twice that for full-wave, where the voltage is not completely smoothed.

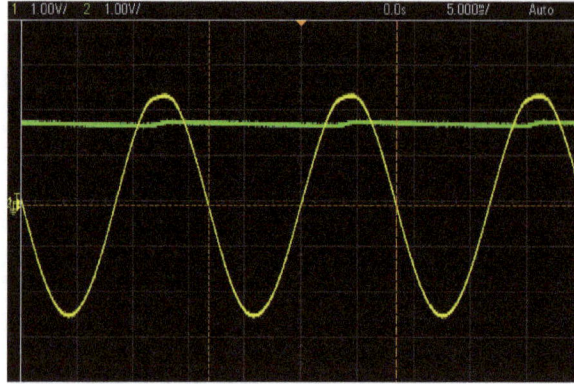

The AC input (yellow) and DC output (green) of a half-wave rectifier with a smoothing capacitor. Note the ripple in the DC signal.

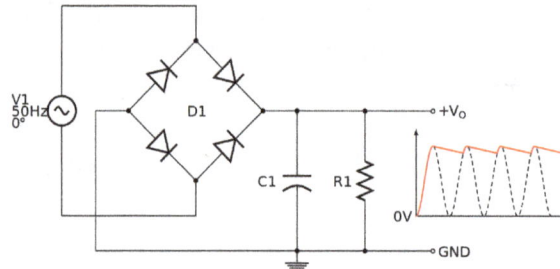

RC-Filter Rectifier: This circuit was designed and simulated using Multisim 8 software.

Sizing of the capacitor represents a tradeoff. For a given load, a larger capacitor reduces ripple but costs more and creates higher peak currents in the transformer secondary and in the supply that feeds it. The peak current is set in principle by the rate of rise of the supply voltage on the rising edge of the incoming sine-wave, but in practice it is reduced by the resistance of the transformer windings. In extreme cases where many rectifiers are loaded onto a power distribution circuit, peak currents may cause difficulty in maintaining a correctly shaped sinusoidal voltage on the ac supply.

To limit ripple to a specified value the required capacitor size is proportional to the load current and inversely proportional to the supply frequency and the number of output peaks of the rectifier per input cycle. The load current and the supply frequency are generally outside the control of the designer of the rectifier system but the number of peaks per input cycle can be affected by the choice of rectifier design.

A half-wave rectifier only gives one peak per cycle, and for this and other reasons is only used in very small power supplies. A full wave rectifier achieves two peaks per cycle, the best possible with a single-phase input. For three-phase inputs a three-phase bridge gives six peaks per cycle. Higher numbers of peaks can be achieved by using transformer networks placed before the rectifier to convert to a higher phase order.

To further reduce ripple, a capacitor-input filter can be used. This complements the reservoir capacitor with a choke (inductor) and a second filter capacitor, so that a steadier DC output can be obtained across the terminals of the filter capacitor. The choke presents a high impedance to the ripple current. For use at power-line frequencies inductors require cores of iron or other magnetic materials, and add weight and size. Their use in power supplies for electronic equipment has therefore dwindled in favour of semiconductor circuits such as voltage regulators.

A more usual alternative to a filter, and essential if the DC load requires very low ripple voltage, is to follow the reservoir capacitor with an active voltage regulator circuit. The reservoir capacitor must be large enough to prevent the troughs of the ripple dropping below the minimum voltage required by the regulator to produce the required output voltage. The regulator serves both to significantly reduce the ripple and to deal with variations in supply and load characteristics. It would be possible to use a smaller reservoir capacitor (these can be large on high-current power supplies) and then apply some filtering as well as the regulator, but this is not a common strategy. The extreme of this approach is to dispense with the reservoir capacitor altogether and put the rectified waveform straight into a choke-input filter. The advantage of this circuit is that the current waveform is smoother and consequently the rectifier no longer has to deal with the current as a large current pulse, but instead the current delivery is spread over the entire cycle. The disadvantage, apart from extra size and weight, is that the voltage output is much lower – approximately the average of an AC half-cycle rather than the peak.

Applications

The primary application of rectifiers is to derive DC power from an AC supply (AC to DC converter). Virtually all electronic devices require DC, so rectifiers are used inside the power supplies of virtually all electronic equipment.

Converting DC power from one voltage to another is much more complicated. One method of DC-to-DC conversion first converts power to AC (using a device called an inverter), then uses a transformer to change the voltage, and finally rectifies power back to DC. A frequency of typically several tens of kilohertz is used, as this requires much smaller inductance than at lower frequencies and obviates the use of heavy, bulky, and expensive iron-cored units.

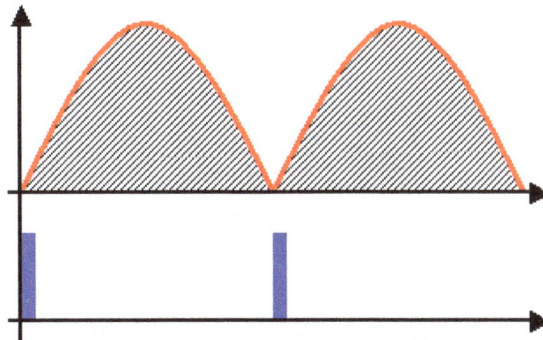

Output voltage of a full-wave rectifier with controlled thyristors

Rectifiers are also used for detection of amplitude modulated radio signals. The signal may be amplified before detection. If not, a very low voltage drop diode or a diode biased with a fixed voltage must be used. When using a rectifier for demodulation the capacitor and load resistance must be carefully matched: too low a capacitance makes the high frequency carrier pass to the output, and too high makes the capacitor just charge and stay charged.

Rectifiers supply polarised voltage for welding. In such circuits control of the output current is required; this is sometimes achieved by replacing some of the diodes in a bridge rectifier with thyristors, effectively diodes whose voltage output can be regulated by switching on and off with phase fired controllers.

Thyristors are used in various classes of railway rolling stock systems so that fine control of the traction motors can be achieved. Gate turn-off thyristors are used to produce alternating current from a DC supply, for example on the Eurostar Trains to power the three-phase traction motors.

Rectification Technologies

Electromechanical

Before about 1905 when tube type rectifiers were developed, power conversion devices were purely electro-mechanical in design. Mechanical rectification systems used some form of rotation or resonant vibration (e.g. vibrators) driven by electromagnets, which operated a switch or commutator to reverse the current.

These mechanical rectifiers were noisy and had high maintenance requirements. The moving parts had friction, which required lubrication and replacement due to wear. Opening mechanical contacts under load resulted in electrical arcs and sparks that heated and eroded the contacts. They also were not able to handle AC frequencies above several thousand cycles per second.

Synchronous Rectifier

To convert alternating into direct current in electric locomotives, a synchronous rectifier may be used. It consists of a synchronous motor driving a set of heavy-duty electrical contacts. The motor spins in time with the AC frequency and periodically reverses the connections to the load at an instant when the sinusoidal current goes through a zero-crossing. The contacts do not have to *switch* a large current, but they must be able to *carry* a large current to supply the locomotive's DC traction motors.

Vibrating Rectifier

A vibrator battery charger from 1922. It produced 6A DC at 6V to charge automobile batteries.

These consisted of a resonant reed, vibrated by an alternating magnetic field created by an AC electromagnet, with contacts that reversed the direction of the current on the negative half cycles.

They were used in low power devices, such as battery chargers, to rectify the low voltage produced by a step-down transformer. Another use was in battery power supplies for portable vacuum tube radios, to provide the high DC voltage for the tubes. These operated as a mechanical version of modern solid state switching inverters, with a transformer to step the battery voltage up, and a set of vibrator contacts on the transformer core, operated by its magnetic field, to repeatedly break the DC battery current to create a pulsing AC to power the transformer. Then a second set of rectifier contacts on the vibrator rectified the high AC voltage from the transformer secondary to DC.

Motor-Generator Set

A small motor-generator set

A *motor-generator set*, or the similar *rotary converter*, is not strictly a rectifier as it does not actually *rectify* current, but rather *generates* DC from an AC source. In an "M-G set", the shaft of an AC motor is mechanically coupled to that of a DC generator. The DC generator produces multiphase alternating currents in its armature windings, which a commutator on the armature shaft converts into a direct current output; or a homopolar generator produces a direct current without the need for a commutator. M-G sets are useful for producing DC for railway traction motors, industrial motors and other high-current applications, and were common in many high-power D.C. uses (for example, carbon-arc lamp projectors for outdoor theaters) before high-power semiconductors became widely available.

Electrolytic

The electrolytic rectifier was a device from the early twentieth century that is no longer used. A home-made version is illustrated in the 1913 book *The Boy Mechanic* but it would only be suitable for use at very low voltages because of the low breakdown voltage and the risk of electric shock. A more complex device of this kind was patented by G. W. Carpenter in 1928 (US Patent 1671970).

When two different metals are suspended in an electrolyte solution, direct current flowing one way through the solution sees less resistance than in the other direction. Electrolytic rectifiers most commonly used an aluminum anode and a lead or steel cathode, suspended in a solution of tri-ammonium ortho-phosphate.

The rectification action is due to a thin coating of aluminum hydroxide on the aluminum electrode, formed by first applying a strong current to the cell to build up the coating. The rectification process is temperature-sensitive, and for best efficiency should not operate above 86 °F (30 °C). There is also a breakdown voltage where the coating is penetrated and the cell is short-circuited. Electrochemical methods are often more fragile than mechanical methods, and can be sensitive to usage variations, which can drastically change or completely disrupt the rectification processes.

Similar electrolytic devices were used as lightning arresters around the same era by suspending many aluminium cones in a tank of tri-ammonium ortho-phosphate solution. Unlike the rectifier above, only aluminium electrodes were used, and used on A.C., there was no polarization and thus no rectifier action, but the chemistry was similar.

The modern electrolytic capacitor, an essential component of most rectifier circuit configurations was also developed from the electrolytic rectifier.

Plasma Type

Mercury-Arc

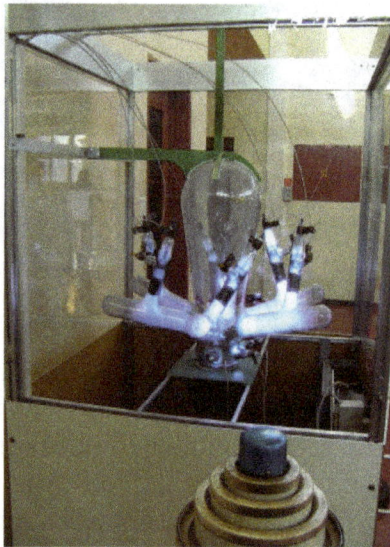

Early 3-phase industrial mercury vapor rectifier tube

150 kV mercury-arc valve at Manitoba Hydro power station, Radisson, Canada converted AC hydropower to DC for transmission to distant cities.

A rectifier used in high-voltage direct current (HVDC) power transmission systems and industrial processing between about 1909 to 1975 is a *mercury-arc rectifier* or *mercury-arc valve*. The device is enclosed in a bulbous glass vessel or large metal tub. One electrode, the cathode, is submerged in a pool of liquid mercury at the bottom of the vessel and one or more high purity graphite electrodes, called anodes, are suspended above the pool. There may be several auxiliary electrodes to aid in starting and maintaining the arc. When an electric arc is established between the cathode pool and suspended anodes, a stream of electrons flows from the cathode to the anodes through the ionized mercury, but not the other way (in principle, this is a higher-power counterpart to flame rectification, which uses the same one-way current transmission properties of the plasma naturally present in a flame).

These devices can be used at power levels of hundreds of kilowatts, and may be built to handle one to six phases of AC current. Mercury-arc rectifiers have been replaced by silicon semiconductor rectifiers and high-power thyristor circuits in the mid 1970s. The most powerful mercury-arc rectifiers ever built were installed in the Manitoba Hydro Nelson River Bipole HVDC project, with a combined rating of more than 1 GW and 450 kV.

Argon Gas Electron Tube

Tungar bulbs from 1917, 2 ampere *(left)* and 6 ampere

The General Electric Tungar rectifier was an argon gas-filled electron tube device with a tungsten filament cathode and a carbon button anode. It operated similarly to the thermionic vacuum tube diode, but the gas in the tube ionized during forward conduction, giving it a much lower forward voltage drop so it could rectify lower voltages. It was used for battery chargers and similar applications from the 1920s until lower-cost metal rectifiers, and later semiconductor diodes, supplanted it. These were made up to a few hundred volts and a few amperes rating, and in some sizes strongly resembled an incandescent lamp with an additional electrode.

The 0Z4 was a gas-filled rectifier tube commonly used in vacuum tube car radios in the 1940s and 1950s. It was a conventional full-wave rectifier tube with two anodes and one cathode, but was unique in that it had no filament (thus the "0" in its type number). The electrodes were shaped such that the reverse breakdown voltage was much higher than the forward breakdown voltage. Once the breakdown voltage was exceeded, the 0Z4 switched to a low-resistance state with a forward voltage drop of about 24 V.

Diode Vacuum Tube (Valve)

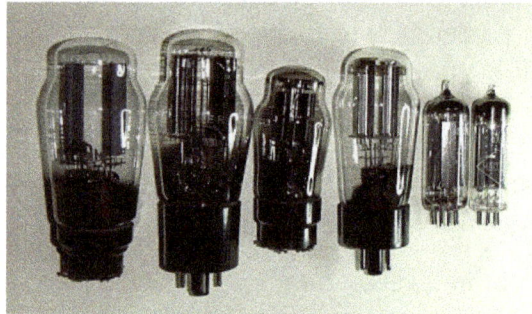

Vacuum tube diodes

The thermionic vacuum tube diode, originally called the Fleming valve, was invented by John Ambrose Fleming in 1904 as a detector for radio waves in radio receivers, and evolved into a general rectifier. It consisted of an evacuated glass bulb with a filament heated by a separate current, and a metal plate anode. The filament emitted electrons by thermionic emission (the Edison effect), discovered by Thomas Edison in 1884, and a positive voltage on the plate caused a current of electrons through the tube from filament to plate. Since only the filament produced electrons, the tube would only conduct current in one direction, allowing the tube to rectify an alternating current.

Vacuum diode rectifiers were widely used in power supplies in vacuum tube consumer electronic products, such as phonographs, radios, and televisions, for example the All American Five radio receiver, to provide the high DC plate voltage needed by other vacuum tubes. "Full-wave" versions with two separate plates were popular because they could be used with a center-tapped transformer to make a full-wave rectifier. Vacuum rectifiers were made for very high voltages, such as the high voltage power supply for the cathode ray tube of television receivers, and the kenotron used for power supply in X-ray equipment. However, compared to modern semiconductor diodes, vacuum rectifiers have high internal resistance due to space charge and therefore high voltage drops, causing high power dissipation and low efficiency. They are rarely able to handle currents exceeding 250 mA owing to the limits of plate power dissipation, and cannot be used for low voltage applications, such as battery chargers. Another limitation of the vacuum tube rectifier is that the heater power supply often requires special arrangements to insulate it from the high voltages of the rectifier circuit.

In musical instrument amplification (especially for electric guitars), the slight delay or "sag" between a signal increase (for instance, when a guitar chord is struck hard and fast) and the corresponding increase in output voltage is a notable effect of tube rectification, and results in compression. The choice between tube rectification and diode rectification is a matter of taste; some amplifiers have both and allow the player to choose.

Solid State

Crystal Detector

The cat's-whisker detector was the earliest type of semiconductor diode. It consisted of a crystal of some semiconducting mineral, usually galena (lead sulfide), with a light springy wire touching its surface. Invented by Jagadish Chandra Bose and developed by G. W. Pickard around 1906, it served as the radio wave rectifier in the first widely used radio receivers, called crystal radios. Its fragility and limited current capability made it unsuitable for power supply applications. It became obsolete around 1920, but later versions served as microwave detectors and mixers in radar receivers during World War 2.

Galena cat's whisker detector

Selenium and Copper Oxide Rectifiers

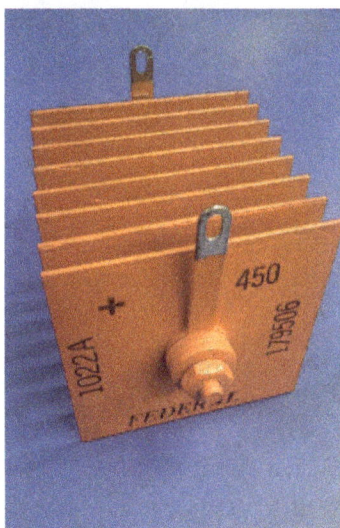

Selenium rectifier

Once common until replaced by more compact and less costly silicon solid-state rectifiers in the 1970s, these units used stacks of metal plates and took advantage of the semiconductor properties of selenium or copper oxide. While selenium rectifiers were lighter in weight and used less power than comparable vacuum tube rectifiers, they had the disadvantage of finite life expectancy, increasing resistance with age, and were only suitable to use at low frequencies. Both selenium and copper oxide rectifiers have somewhat better tolerance of momentary voltage transients than silicon rectifiers.

Typically these rectifiers were made up of stacks of metal plates or washers, held together by a central bolt, with the number of stacks determined by voltage; each cell was rated for about 20 V. An automotive battery charger rectifier might have only one cell: the high-voltage power supply for a vacuum tube might have dozens of stacked plates. Current density in an air-cooled selenium stack was about 600 mA per square inch of active area (about 90 mA per square centimeter).

Silicon and Germanium Diodes

Diodes

In the modern world, silicon diodes are the most widely used rectifiers for lower voltages and powers, and have largely replaced earlier germanium diodes. For very high voltages and powers, the added need for controllability has in practice led to replacing simple silicon diodes with high-power thyristors and their newer actively gate-controlled cousins.

High Power: Thyristors (SCRs) and Newer Silicon-Based Voltage Sourced Converters

Two of three high-power thyristor valve stacks used for long distance transmission of power from Manitoba Hydro dams. Compare with mercury-arc system from the same dam-site, above.

In high-power applications, from 1975 to 2000, most mercury valve arc-rectifiers were replaced by stacks of very high power thyristors, silicon devices with two extra layers of semiconductor, in comparison to a simple diode.

In medium-power transmission applications, even more complex and sophisticated voltage sourced converter (VSC) silicon semiconductor rectifier systems, such as insulated gate bipolar transistors (IGBT) and gate turn-off thyristors (GTO), have made smaller high voltage DC power transmission systems economical. All of these devices function as rectifiers.

As of 2009 it was expected that these high-power silicon "self-commutating switches", in particular IGBTs and a variant thyristor (related to the GTO) called the integrated gate-commutated thyristor (IGCT), would be scaled-up in power rating to the point that they would eventually replace simple thyristor-based AC rectification systems for the highest power-transmission DC applications.

Current Research

A major area of research is to develop higher frequency rectifiers, that can rectify into terahertz and light frequencies. These devices are used in optical heterodyne detection, which has myriad applications in optical fiber communication and atomic clocks. Another prospective application for such devices is to directly rectify light waves picked up by tiny antenna, called nantennas, to produce DC electric power. It is thought that arrays of nantennas could be a more efficient means of producing solar power than solar cells.

A related area of research is to develop smaller rectifiers, because a smaller device has a higher cutoff frequency. Research projects are attempting to develop a unimolecular rectifier, a single organic molecule that would function as a rectifier.

References

- Bose, Bimal K. (2006). Power Electronics and Motor Drives : Advances and Trends. Amsterdam: Academic. p. 126. ISBN 978-0-12-088405-6.

- Stephen Sangwine (2 March 2007). Electronic Components and Technology, Third Edition. CRC Press. p. 73. ISBN 978-1-4200-0768-8.

- Williams, B. W. (1992). "Chapter 11". Power electronics : devices, drivers and applications (2nd ed.). Basingstoke: Macmillan. ISBN 978-0-333-57351-8.

- Wendy Middleton, Mac E. Van Valkenburg (eds), Reference Data for Engineers: Radio, Electronics, Computer, and Communications, p. 14. 13, Newnes, 2002 ISBN 0-7506-7291-9.

- Arrillaga, Jos; Liu, Yonghe H; Watson, Neville R; Murray, Nicholas J. Self-Commutating Converters for High Power Applications. John Wiley & Sons. ISBN 978-0-470-68212-8.

- Kimbark, Edward Wilson (1971). Direct current transmission. (4. printing. ed.). New York: Wiley-Interscience. p. 508. ISBN 978-0-471-47580-4.

- Ed Brorein (2012-05-16). "Watt's Up?: What Is Old is New Again: Soft-Switching and Synchronous Rectification in Vintage Automobile Radios". Keysight Technologies: Watt's Up?. Retrieved 2016-01-19.

- Jeff Barrow of Integrated Device Technology, Inc. (21 November 2011). "Understand and reduce DC/DC switching-converter ground noise". Eetimes.com. Retrieved 18 January 2016.

- Majumder, Ritwik; Ghosh, Arindam; Ledwich, Gerard F.; Zare, Firuz (2008). "Control of Parallel Converters for Load Sharing with Seamless Transfer between Grid Connected and Islanded Modes". eprints.qut.edu.au. Retrieved 2016-01-19.

- Hendrik Rissik (1941). Mercury-arc current convertors [sic] : an introduction to the theory and practice of vapour-arc discharge devices and to the study of rectification phenomena. Sir I. Pitman & sons, ltd. Retrieved 8 January 2013.

- Hawkins, Nehemiah (1914). "54. Rectifiers". Hawkins Electrical Guide: Principles of electricity, magnetism, induction, experiments, dynamo. New York: T. Audel. Retrieved 8 January 2013.

- American Technical Society (1920). Cyclopedia of applied electricity. 2. American technical society. p. 487. Retrieved 8 January 2013.

- "A SIDE-BY-SIDE COMPARISON OF MICRO AND CENTRAL INVERTERS IN SHADED AND UNSHADED CONDITIONS" (PDF). Retrieved 27 August 2013.

- Benanti, Travis L.; Venkataraman, D. (25 April 2005). "Organic Solar Cells: An Overview Focusing on Active Layer Morphology" (PDF). Photosynthesis Research. 87 (1): 77. doi:10.1007/s11120-005-6397-9. Retrieved 27 August 2013.

- Swamy, Mahesh; Kume, Tsuneo (Dec 16, 2010). "Present State and Futuristic Vision of Motor Drive Technology" (PDF). Power Transmission Engineering. www.powertransmission.com. Retrieved Apr 2012.

- Pakaste, Risto; et al. (Feb 1999). "Experience with Azipod propulsion systems on board marine vessels" (PDF). Retrieved 28 April 2012.

5

Applications of Power Electronics

Switched-mode power supply is a method of regulating yield voltage or current by switching elements of the circuit thereby increasing efficiency. The chapter explores fluorescent lamps, battery charger and switched-mode power supply and provides the principles of operation and applications of each. The aspects elucidated in this chapter are of vital importance, and provide a better understanding of power electronics.

Switched-mode Power Supply

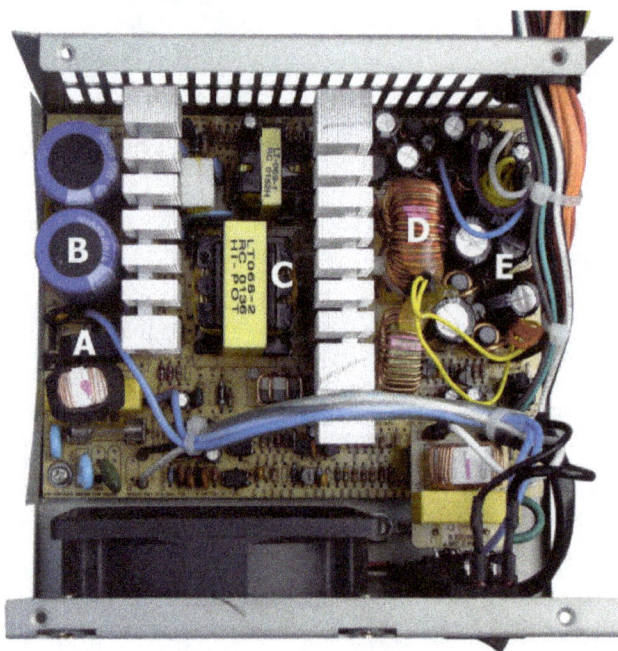

Interior view of an ATX SMPS: below

A: input EMI filtering and bridge rectifier;

B: input filter capacitors;

Between B and C: primary side heat sink;

C: transformer;

Between C and D: secondary side heat sink;

D: output filter coil;

E: output filter capacitors.

The coil and large yellow capacitor below E are additional input filtering components that are mounted directly on the power input connector and are not part of the main circuit board.

An adjustable switched-mode power supply for laboratory use

A switched-mode power supply (switching-mode power supply, switch-mode power supply, switched power supply, SMPS, or switcher) is an electronic power supply that incorporates a switching regulator to convert electrical power efficiently. Like other power supplies, an SMPS transfers power from a DC or AC source (often mains power), to DC loads, such as a personal computer, while converting voltage and current characteristics. Unlike a linear power supply, the pass transistor of a switching-mode supply continually switches between low-dissipation, full-on and full-off states, and spends very little time in the high dissipation transitions, which minimizes wasted energy. Ideally, a switched-mode power supply dissipates no power. Voltage regulation is achieved by varying the ratio of on-to-off time. In contrast, a linear power supply regulates the output voltage by continually dissipating power in the pass transistor. This higher power conversion efficiency is an important advantage of a switched-mode power supply. Switched-mode power supplies may also be substantially smaller and lighter than a linear supply due to the smaller transformer size and weight.

Switching regulators are used as replacements for linear regulators when higher efficiency, smaller size or lighter weight are required. They are, however, more complicated; their switching currents can cause electrical noise problems if not carefully suppressed, and simple designs may have a poor power factor.

History

1836

Induction coils use switches to generate high voltages.

1910

An inductive discharge ignition system invented by Charles F. Kettering and his company Dayton Engineering Laboratories Company (Delco) goes into production for Cadillac. The Kettering ignition system is a mechanically-switched version of a flyback boost converter; the transformer is the ignition coil. Variations of this ignition system were used in all non-diesel internal combustion engines until the 1960s when it was displaced with capacitive discharge ignition systems.

1926

On 23 June, British inventor Philip Ray Coursey applies for a patent in his country and

United States, for his "Electrical Condenser". The patent mentions high frequency welding and furnaces, among other uses.

ca 1936

Car radios used electromechanical vibrators to transform the 6 V battery supply to a suitable B+ voltage for the vacuum tubes.

1959

Transistor oscillation and rectifying converter power supply system U.S. Patent 3,040,271 is filed by Joseph E. Murphy and Francis J. Starzec, from General Motors Company

1970

Tektronix starts using High-Efficiency Power Supply in its 7000-series oscilloscopes produced from about 1970 to 1995.

1972

HP-35, Hewlett-Packard's first pocket calculator, is introduced with transistor switching power supply for light-emitting diodes, clocks, timing, ROM, and registers.

1973

Xerox uses switching power supplies in the Alto minicomputer

1977

Apple II is designed with a switching mode power supply. *"Rod Holt was brought in as product engineer and there were several flaws in Apple II that were never publicized. One thing Holt has to his credit is that he created the switching power supply that allowed us to do a very lightweight computer"*.

1980

The HP8662A 10 kHz – 1.28 GHz synthesized signal generator went with a switched mode power supply.

Explanation

A linear regulator provides the desired output voltage by dissipating excess power in ohmic losses (e.g., in a resistor or in the collector–emitter region of a pass transistor in its active mode). A linear regulator regulates either output voltage or current by dissipating the excess electric power in the form of heat, and hence its maximum power efficiency is voltage-out/voltage-in since the volt difference is wasted.

In contrast, a switched-mode power supply regulates either output voltage or current by switching ideal storage elements, like inductors and capacitors, into and out of different electrical configurations. Ideal switching elements (e.g., transistors operated outside of their active mode) have no resistance when "closed" and carry no current when "open", and so the converters can theoretically operate with 100% efficiency (i.e., all input power is delivered to the load; no power is wasted as dissipated heat).

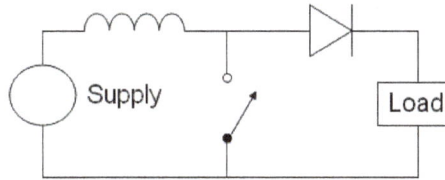

The basic schematic of a boost converter.

For example, if a DC source, an inductor, a switch, and the corresponding electrical ground are placed in series and the switch is driven by a square wave, the peak-to-peak voltage of the waveform measured across the switch can exceed the input voltage from the DC source. This is because the inductor responds to changes in current by inducing its own voltage to counter the change in current, and this voltage adds to the source voltage while the switch is open. If a diode-and-capacitor combination is placed in parallel to the switch, the peak voltage can be stored in the capacitor, and the capacitor can be used as a DC source with an output voltage greater than the DC voltage driving the circuit. This boost converter acts like a step-up transformer for DC signals. A buck–boost converter works in a similar manner, but yields an output voltage which is opposite in polarity to the input voltage. Other buck circuits exist to boost the average output current with a reduction of voltage.

In a SMPS, the output current flow depends on the input power signal, the storage elements and circuit topologies used, and also on the pattern used (e.g., pulse-width modulation with an adjustable duty cycle) to drive the switching elements. The spectral density of these switching waveforms has energy concentrated at relatively high frequencies. As such, switching transients and ripple introduced onto the output waveforms can be filtered with a small LC filter.

Advantages and Disadvantages

The main advantage of the switching power supply is greater efficiency than linear regulators because the switching transistor dissipates little power when acting as a switch.

Other advantages include smaller size and lighter weight from the elimination of heavy line-frequency transformers, and comparable heat generation. Standby power loss is often much less than transformers.

Disadvantages include greater complexity, the generation of high-amplitude, high-frequency energy that the low-pass filter must block to avoid electromagnetic interference (EMI), a ripple voltage at the switching frequency and the harmonic frequencies thereof.

Very low cost SMPSs may couple electrical switching noise back onto the mains power line, causing interference with A/V equipment connected to the same phase. Non-power-factor-corrected SMPSs also cause harmonic distortion.

SMPS and Linear Power Supply Comparison

There are two main types of regulated power supplies available: SMPS and linear. The following table compares linear regulated and unregulated AC-to-DC supplies with switching regulators in general:

Comparison of a linear power supply and a switched-mode power supply			
	Linear power supply	Switching power supply	Notes
Size and weight	Heatsinks for high power linear regulators add size and weight. Transformers, if used, are large due to low operating frequency (mains power frequency is at 50 or 60 Hz); otherwise can be compact due to low component count.	Smaller transformer (if used; else inductor) due to higher operating frequency (typically 50 kHz – 1 MHz). Size and weight of adequate RF shielding may be significant.	A transformer's power handling capacity of given size and weight increases with frequency provided that hysteresis losses can be kept down. Therefore, higher operating frequency means either a higher capacity or smaller transformer.
Output voltage	With transformer used, any voltages available; if transformerless, limited to what can be achieved with a voltage doubler. If unregulated, voltage varies significantly with load.	Any voltages available, limited only by transistor breakdown voltages in many circuits. Voltage varies little with load.	A SMPS can usually cope with wider variation of input before the output voltage changes.
Efficiency, heat, and power dissipation	If regulated: efficiency largely depends on voltage difference between input and output; output voltage is regulated by dissipating excess power as heat resulting in a typical efficiency of 30–40%. If unregulated, transformer iron and copper losses may be the only significant sources of inefficiency.	Output is regulated using duty cycle control; the transistors are switched fully on or fully off, so very little resistive losses between input and the load. The only heat generated is in the non-ideal aspects of the components and quiescent current in the control circuitry.	Switching losses in the transistors (especially in the short part of each cycle when the device is partially on), on-resistance of the switching transistors, equivalent series resistance in the inductor and capacitors, and core losses in the inductor, and rectifier voltage drop contribute to a typical efficiency of 60–70%. However, by optimizing SMPS design (such as choosing the optimal switching frequency, avoiding saturation of inductors, and active rectification), the amount of power loss and heat can be minimized; a good design can have an efficiency of 95%.

Complexity	Unregulated may be simply a diode and capacitor; regulated has a voltage-regulating circuit and a noise-filtering capacitor; usually a simpler circuit (and simpler feedback loop stability criteria) than switched-mode circuits.	Consists of a controller IC, one or several power transistors and diodes as well as a power transformer, inductors, and filter capacitors. Some design complexities present (reducing noise/interference; extra limitations on maximum ratings of transistors at high switching speeds) not found in linear regulator circuits.	In switched-mode mains (AC-to-DC) supplies, multiple voltages can be generated by one transformer core, but that can introduce design/use complications: for example it may place minimum output current restrictions on one output. For this SMPSs have to use duty cycle control. One of the outputs has to be chosen to feed the voltage regulation feedback loop (usually 3.3 V or 5 V loads are more fussy about their supply voltages than the 12 V loads, so this drives the decision as to which feeds the feedback loop. The other outputs usually track the regulated one pretty well). Both need a careful selection of their transformers. Due to the high operating frequencies in SMPSs, the stray inductance and capacitance of the printed circuit board traces become important.
Radio frequency interference	Mild high-frequency interference may be generated by AC rectifier diodes under heavy current loading, while most other supply types produce no high-frequency interference. Some mains hum induction into unshielded cables, problematical for low-signal audio.	EMI/RFI produced due to the current being switched on and off sharply. Therefore, EMI filters and RF shielding are needed to reduce the disruptive interference.	Long wires between the components may reduce the high frequency filter efficiency provided by the capacitors at the inlet and outlet. Stable switching frequency may be important.
Electronic noise at the output terminals	Unregulated PSUs may have a little AC ripple superimposed upon the DC component at twice mains frequency (100–120 Hz). It can cause audible mains hum in audio equipment, brightness ripples or banded distortions in analog security cameras.	Noisier due to the switching frequency of the SMPS. An unfiltered output may cause glitches in digital circuits or noise in audio circuits.	This can be suppressed with capacitors and other filtering circuitry in the output stage. With a switched mode PSU the switching frequency can be chosen to keep the noise out of the circuits working frequency band (e.g., for audio systems above the range of human hearing)

Electronic noise at the input terminals	Causes harmonic distortion to the input AC, but relatively little or no high frequency noise.	Very low cost SMPS may couple electrical switching noise back onto the mains power line, causing interference with A/V equipment connected to the same phase. Non power-factor-corrected SMPSs also cause harmonic distortion.	This can be prevented if a (properly earthed) EMI/RFI filter is connected between the input terminals and the bridge rectifier.
Acoustic noise	Faint, usually inaudible mains hum, usually due to vibration of windings in the transformer or magnetostriction.	Usually inaudible to most humans, unless they have a fan or are unloaded/malfunctioning, or use a switching frequency within the audio range, or the laminations of the coil vibrate at a subharmonic of the operating frequency.	The operating frequency of an unloaded SMPS is sometimes in the audible human range, and may sound subjectively quite loud for people whose hearing is very sensitive to the relevant frequency range.
Power factor	Low for a regulated supply because current is drawn from the mains at the peaks of the voltage sinusoid, unless a choke-input or resistor-input circuit follows the rectifier (now rare).	Ranging from very low to medium since a simple SMPS without PFC draws current spikes at the peaks of the AC sinusoid.	Active/passive power factor correction in the SMPS can offset this problem and are even required by some electric regulation authorities, particularly in the EU. The internal resistance of low-power transformers in linear power supplies usually limits the peak current each cycle and thus gives a better power factor than many switched-mode power supplies that directly rectify the mains with little series resistance.
Inrush current	Large current when mains-powered linear power supply equipment is switched on until magnetic flux of transformer stabilises and capacitors charge completely, unless a slow-start circuit is used.	Extremely large peak "in-rush" surge current limited only by the impedance of the input supply and any series resistance to the filter capacitors.	Empty filter capacitors initially draw large amounts of current as they charge up, with larger capacitors drawing larger amounts of peak current. Being many times above the normal operating current, this greatly stresses components subject to the surge, complicates fuse selection to avoid nuisance blowing and may cause problems with equipment employing overcurrent protection such as uninterruptible power supplies. Mitigated by use of a suitable soft-start circuit or series resistor.

Risk of electric shock	Supplies with transformers isolate the incoming power supply from the powered device and so allow metalwork of the enclosure to be grounded safely. Dangerous if primary/secondary insulation breaks down, unlikely with reasonable design. Transformerless mains-operated supply dangerous. In both linear and switch-mode the mains, and possibly the output voltages, are hazardous and must be well-isolated.	Common rail of equipment (including casing) is energized to half the mains voltage, but at high impedance, unless equipment is earthed/grounded or doesn't contain EMI/RFI filtering at the input terminals.	Due to regulations concerning EMI/RFI radiation, many SMPS contain EMI/RFI filtering at the input stage before the bridge rectifier consisting of capacitors and inductors. Two capacitors are connected in series with the Live and Neutral rails with the Earth connection in between the two capacitors. This forms a capacitive divider that energizes the common rail at half mains voltage. Its high impedance current source can provide a tingling or a 'bite' to the operator or can be exploited to light an Earth Fault LED. However, this current may cause nuisance tripping on the most sensitive residual-current devices.
Risk of equipment damage	Very low, unless a short occurs between the primary and secondary windings or the regulator fails by shorting internally.	Can fail so as to make output voltage very high. Stress on capacitors may cause them to explode. Can in some cases destroy input stages in amplifiers if floating voltage exceeds transistor base-emitter breakdown voltage, causing the transistor's gain to drop and noise levels to increase. Mitigated by good failsafe design. Failure of a component in the SMPS itself can cause further damage to other PSU components; can be difficult to troubleshoot.	The floating voltage is caused by capacitors bridging the primary and secondary sides of the power supply. Connection to earthed equipment will cause a momentary (and potentially destructive) spike in current at the connector as the voltage at the secondary side of the capacitor equalizes to earth potential.

Theory of Operation

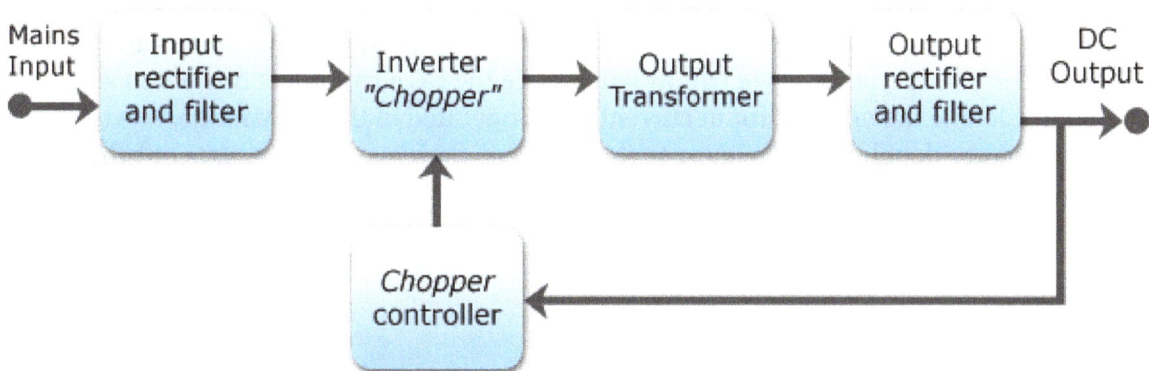

Block diagram of a mains operated AC/DC SMPS with output voltage regulation

Input Rectifier Stage

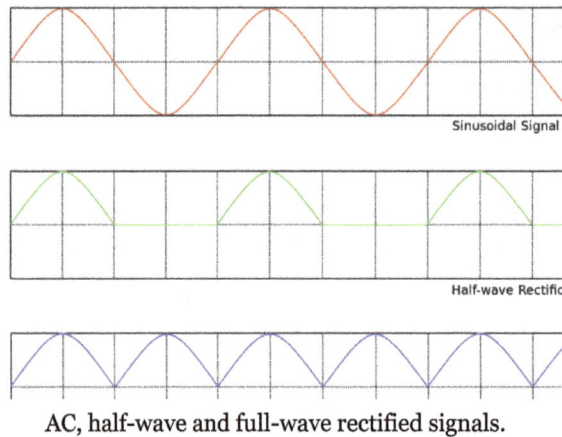

AC, half-wave and full-wave rectified signals.

If the SMPS has an AC input, then the first stage is to convert the input to DC. This is called *rectification*. A SMPS with a DC input does not require this stage. In some power supplies (mostly computer ATX power supplies), the rectifier circuit can be configured as a voltage doubler by the addition of a switch operated either manually or automatically. This feature permits operation from power sources that are normally at 115 V or at 230 V. The rectifier produces an unregulated DC voltage which is then sent to a large filter capacitor. The current drawn from the mains supply by this rectifier circuit occurs in short pulses around the AC voltage peaks. These pulses have significant high frequency energy which reduces the power factor. To correct for this, many newer SMPS will use a special PFC circuit to make the input current follow the sinusoidal shape of the AC input voltage, correcting the power factor. Power supplies that use Active PFC usually are auto-ranging, supporting input voltages from ~100 VAC – 250 VAC, with no input voltage selector switch.

An SMPS designed for AC input can usually be run from a DC supply, because the DC would pass through the rectifier unchanged. If the power supply is designed for 115 VAC and has no voltage selector switch, the required DC voltage would be 163 VDC (115 × √2). This type of use may be harmful to the rectifier stage, however, as it will only use half of diodes in the rectifier for the full load. This could possibly result in overheating of these components, causing them to fail prematurely. On the other hand, if the power supply has a voltage selector switch, based on the Delon circuit, for 115/230V (computer ATX power supplies typically are in this category), the selector switch would have to be put in the 230 V position, and the required voltage would be 325 VDC (230 × √2). The diodes in this type of power supply will handle the DC current just fine because they are rated to handle double the nominal input current when operated in the 115 V mode, due to the operation of the voltage doubler. This is because the doubler, when in operation, uses only half of the bridge rectifier and runs twice as much current through it. It is uncertain how an Auto-ranging/Active-PFC type power supply would react to being powered by DC.

Inverter Stage

The inverter stage converts DC, whether directly from the input or from the rectifier stage described above, to AC by running it through a power oscillator, whose output transformer is very small with few windings at a frequency of tens or hundreds of kilohertz. The frequency is usually chosen to be above 20 kHz, to make it inaudible to humans. The switching is implemented as a multistage (to achieve high gain) MOSFET amplifier. MOSFETs are a type of transistor with a low on-resistance and a high current-handling capacity.

Voltage Converter and Output Rectifier

If the output is required to be isolated from the input, as is usually the case in mains power supplies, the inverted AC is used to drive the primary winding of a high-frequency transformer. This converts the voltage up or down to the required output level on its secondary winding. The output transformer in the block diagram serves this purpose.

If a DC output is required, the AC output from the transformer is rectified. For output voltages above ten volts or so, ordinary silicon diodes are commonly used. For lower voltages, Schottky diodes are commonly used as the rectifier elements; they have the advantages of faster recovery times than silicon diodes (allowing low-loss operation at higher frequencies) and a lower voltage drop when conducting. For even lower output voltages, MOSFETs may be used as synchronous rectifiers; compared to Schottky diodes, these have even lower conducting state voltage drops.

The rectified output is then smoothed by a filter consisting of inductors and capacitors. For higher switching frequencies, components with lower capacitance and inductance are needed.

Simpler, non-isolated power supplies contain an inductor instead of a transformer. This type includes *boost converters*, *buck converters*, and the *buck-boost converters*. These belong to the simplest class of single input, single output converters which use one inductor and one active switch. The buck converter reduces the input voltage in direct proportion to the ratio of conductive time to the total switching period, called the duty cycle. For example an ideal buck converter with a 10 V input operating at a 50% duty cycle will produce an average output voltage of 5 V. A feedback control loop is employed to regulate the output voltage by varying the duty cycle to compensate for variations in input voltage. The output voltage of a boost converter is always greater than the input voltage and the buck-boost output voltage is inverted but can be greater than, equal to, or less than the magnitude of its input voltage. There are many variations and extensions to this class of converters but these three form the basis of almost all isolated and non-isolated DC to DC converters. By adding a second inductor the Ćuk and SEPIC converters can be implemented, or, by adding additional active switches, various bridge converters can be realized.

Other types of SMPSs use a capacitor-diode voltage multiplier instead of inductors and transformers. These are mostly used for generating high voltages at low currents (*Cockcroft-Walton generator*). The low voltage variant is called charge pump.

Regulation

A feedback circuit monitors the output voltage and compares it with a reference voltage, as shown in the block diagram above. Depending on design and safety requirements, the controller may contain an isolation mechanism (such as an opto-coupler) to isolate it from the DC output. Switching supplies in computers, TVs and VCRs have these opto-couplers to tightly control the output voltage.

This charger for a small device such as a mobile phone is a simple off-line switching power supply with a European plug.

Open-loop regulators do not have a feedback circuit. Instead, they rely on feeding a constant voltage to the input of the transformer or inductor, and assume that the output will be correct. Regulated designs compensate for the impedance of the transformer or coil. Monopolar designs also compensate for the magnetic hysteresis of the core.

The feedback circuit needs power to run before it can generate power, so an additional non-switching power-supply for stand-by is added.

Transformer Design

Any switched-mode power supply that gets its power from an AC power line (called an "off-line" converter) requires a transformer for galvanic isolation. Some DC-to-DC converters may also include a transformer, although isolation may not be critical in these cases. SMPS transformers run at high frequency. Most of the cost savings (and space savings) in off-line power supplies result from the smaller size of the high frequency transformer compared to the 50/60 Hz transformers formerly used. There are additional design tradeoffs.

The terminal voltage of a transformer is proportional to the product of the core area, magnetic flux, and frequency. By using a much higher frequency, the core area (and so the mass of the core) can be greatly reduced. However, core losses increase at higher frequencies. Cores generally use ferrite material which has a low loss at the high frequencies and high flux densities used. The laminated iron cores of lower-frequency (<400 Hz) transformers would be unacceptably lossy at switching frequencies of a few kilohertz. Also, more energy is lost during transitions of the switching semiconductor at higher frequencies. Furthermore, more attention to the physical layout of the circuit board is required as parasitics become more significant, and the amount of electromagnetic interference will be more pronounced.

Copper Loss

At low frequencies (such as the line frequency of 50 or 60 Hz), designers can usually ignore the skin effect. For these frequencies, the skin effect is only significant when the conductors are large, more than 0.3 inches (7.6 mm) in diameter.

Switching power supplies must pay more attention to the skin effect because it is a source of power loss. At 500 kHz, the skin depth in copper is about 0.003 inches (0.076 mm) – a dimension smaller than the typical wires used in a power supply. The effective resistance of conductors increases, because current concentrates near the surface of the conductor and the inner portion carries less current than at low frequencies.

The skin effect is exacerbated by the harmonics present in the high speed PWM switching waveforms. The appropriate skin depth is not just the depth at the fundamental, but also the skin depths at the harmonics.

In addition to the skin effect, there is also a proximity effect, which is another source of power loss.

Power Factor

Simple off-line switched mode power supplies incorporate a simple full-wave rectifier connected to a large energy storing capacitor. Such SMPSs draw current from the AC line in short pulses when the mains instantaneous voltage exceeds the voltage across this capacitor. During the remaining portion of the AC cycle the capacitor provides energy to the power supply.

As a result, the input current of such basic switched mode power supplies has high harmonic content and relatively low power factor. This creates extra load on utility lines, increases heating of building wiring, the utility transformers, and standard AC electric motors, and may cause stability problems in some applications such as in emergency generator systems or aircraft generators. Harmonics can be removed by filtering, but the filters are expensive. Unlike displacement power factor created by linear inductive or capacitive loads, this distortion cannot be corrected by addition of a single linear component. Additional circuits are required to counteract the effect of the brief current pulses. Putting a current regulated boost chopper stage after the off-line rectifier (to charge the storage capacitor) can correct the power factor, but increases the complexity and cost.

In 2001, the European Union put into effect the standard IEC/EN61000-3-2 to set limits on the harmonics of the AC input current up to the 40th harmonic for equipment above 75 W. The standard defines four classes of equipment depending on its type and current waveform. The most rigorous limits (class D) are established for personal computers, computer monitors, and TV receivers. To comply with these requirements, modern switched-mode power supplies normally include an additional power factor correction (PFC) stage.

Types

Switched-mode power supplies can be classified according to the circuit topology. The most important distinction is between isolated converters and non-isolated ones.

Non-Isolated Topologies

Non-isolated converters are simplest, with the three basic types using a single inductor for energy storage. In the voltage relation column, D is the duty cycle of the converter, and can vary from 0 to 1. The input voltage (V_1) is assumed to be greater than zero; if it is negative, for consistency, negate the output voltage (V_2).

Type	Typical Power [W]	Relative cost	Energy storage	Voltage relation	Features
Buck	0–1,000	1.0	Single inductor	0 ≤ Out ≤ In,	Current is continuous at output.
Boost	0–5,000	1.0	Single inductor	Out ≥ In,	Current is continuous at input.
Buck-boost	0–150	1.0	Single inductor	Out ≤ 0,	Current is dis-continuous at both input and output.
Split-pi (or, boost-buck)	0–4,500	>2.0	Two inductors and three capacitors	Up or down	Bidirectional power control; in or out
Ćuk			Capacitor and two inductors	Any inverted,	Current is continuous at input and output
SEPIC			Capacitor and two inductors	Any,	Current is continuous at input
Zeta			Capacitor and two inductors	Any,	Current is continuous at output
Charge pump / Switched capacitor			Capacitors only		No magnetic energy storage is needed to achieve conversion, however high efficiency power processing is normally limited to a discrete set of conversion ratios.

When equipment is human-accessible, voltage and power limits of <=42.4 V peak/60 V dc and 250 VA apply for Safety Certification (UL, CSA, VDE approval).

The buck, boost, and buck-boost topologies are all strongly related. Input, output and ground come together at one point. One of the three passes through an inductor on the way, while the other two pass through switches. One of the two switches must be active (e.g., a transistor), while the other can be a diode. Sometimes, the topology can be changed simply by re-labeling the connections. A 12 V input, 5 V output buck converter can be converted to a 7 V input, –5 V output buck-boost by grounding the *output* and taking the output from the *ground* pin.

Likewise, SEPIC and Zeta converters are both minor rearrangements of the Ćuk converter.

The Neutral Point Clamped (NPC) topology is used in power supplies and active filters and is mentioned here for completeness.

Switchers become less efficient as duty cycles become extremely short. For large voltage changes, a transformer (isolated) topology may be better.

Isolated Topologies

All isolated topologies include a transformer, and thus can produce an output of higher or lower voltage than the input by adjusting the turns ratio. For some topologies, multiple windings can be placed on the transformer to produce multiple output voltages. Some converters use the transformer for energy storage, while others use a separate inductor.

Type	Power [W]	Relative cost	Input range [V]	Energy storage	Features
Flyback	0–250	1.0	5–600	Mutual Inductors	Isolated form of the buck-boost converter.[1]
Ringing choke converter (RCC)	0–150	1.0	5–600	Transformer	Low-cost self-oscillating flyback variant.
Half-forward	0–250	1.2	5–500	Inductor	
Forward[2]	100-200		60–200	Inductor	Isolated form of buck converter
Resonant forward	0–60	1.0	60–400	Inductor and capacitor	Single rail input, unregulated output, high efficiency, low EMI.
Push-pull	100–1,000	1.75	50–1,000	Inductor	
Half-bridge	0–2,000	1.9	50–1,000	Inductor	
Full-bridge	400–5,000	>2.0	50–1,000	Inductor	Very efficient use of transformer, used for highest powers.
Resonant, zero voltage switched	>1,000	>2.0		Inductor and capacitor	
Isolated Ćuk				Two capacitors and two inductors	

Zero voltage switched mode power supplies require only small heatsinks as little energy is lost as heat. This allows them to be small. This ZVS can deliver more than 1 kilowatt. Transformer is not shown.

- Flyback converter logarithmic control loop behavior might be harder to control than other types.

- The forward converter has several variants, varying in how the transformer is "reset" to zero magnetic flux every cycle.

Quasi-Resonant Zero-Current/Zero-Voltage Switch

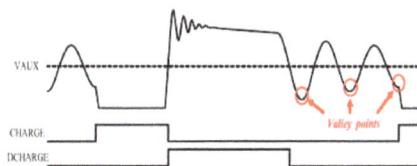

Quasi-resonant switching switches when the voltage is at a minimum and a valley is detected

In a quasi-resonant zero-current/zero-voltage switch (ZCS/ZVS) "each switch cycle delivers a quantized 'packet' of energy to the converter output, and switch turn-on and turn-off occurs at zero current and voltage, resulting in an essentially lossless switch." Quasi-resonant switching, also known as *valley switching*, reduces EMI in the power supply by two methods:

- By switching the bipolar switch when the voltage is at a minimum (in the valley) to minimize the hard switching effect that causes EMI.

- By switching when a valley is detected, rather than at a fixed frequency, introduces a natural frequency jitter that spreads the RF emissions spectrum and reduces overall EMI.

Efficiency and EMI

Higher input voltage and synchronous rectification mode makes the conversion process more efficient. The power consumption of the controller also has to be taken into account. Higher switching frequency allows component sizes to be shrunk, but can produce more RFI. A resonant forward converter produces the lowest EMI of any SMPS approach because it uses a soft-switching resonant waveform compared with conventional hard switching.

Failure Modes

For failure in switching components, circuit board and so on read the failure modes of electronics article.

Power supplies which use capacitors suffering from the capacitor plague may experience premature failure when the capacitance drops to 4% of the original value. This usually causes the switching semiconductor to fail in a conductive way. That may expose connected loads to the full input volt and current, and precipitate wild oscillations in output.

Failure of the switching transistor is common. Due to the large switching voltages this transistor must handle (around 325 V for a 230 V_{AC} mains supply), these transistors often short out, in turn immediately blowing the main internal power fuse.

Precautions

The main filter capacitor will often store up to 325 volts long after the power cord has been removed from the wall. Not all power supplies contain a small "bleeder" resistor to slowly discharge this capacitor. Any contact with this capacitor may result in a severe electrical shock.

The primary and secondary sides may be connected with a capacitor to reduce EMI and compensate for various capacitive couplings in the converter circuit, where the transformer is one. This may result in electric shock in some cases. The current flowing from line or neutral through a 2 kΩ resistor to any accessible part must, according to IEC 60950, be less than 250 μA for IT equipment.

Applications

Switched-mode power supply units (PSUs) in domestic products such as personal computers often have universal inputs, meaning that they can accept power from mains supplies throughout the world, although a manual voltage range switch may be required. Switch-mode power supplies can tolerate a wide range of power frequencies and voltages.

Switched mode mobile phone charger

A 450 Watt SMPS for use in personal computers with the power input, fan, and output cords visible

Due to their high volumes mobile phone chargers have always been particularly cost sensitive. The first chargers were linear power supplies but they quickly moved to the cost effective ringing choke converter (RCC) SMPS topology, when new levels of efficiency were required. Recently, the demand for even lower no-load power requirements in the application has meant that flyback topology is being used more widely; primary side sensing flyback controllers are also helping to cut the bill of materials (BOM) by removing secondary-side sensing components such as optocouplers.

Switched-mode power supplies are used for DC to DC conversion as well. In automobiles where heavy vehicles use a nominal 24 V_{DC} cranking supply, 12V for accessories may be furnished through a DC/DC switch-mode supply. This has the advantage over tapping the battery at the 12V position (using half the cells) that all the 12V load is evenly divided over all cells of the 24V battery. In industrial settings such as telecommunications racks, bulk power may be distributed at a low DC voltage (from a battery back up system, for example) and individual equipment items will have DC/DC switched-mode converters to supply whatever voltages are needed.

Terminology

The term switchmode was widely used until Motorola claimed ownership of the trademark SWITCHMODE, for products aimed at the switching-mode power supply market, and started to enforce their trademark. *Switching-mode power supply, switching power supply,* and *switching regulator* refer to this type of power supply.

Fluorescent Lamp

A fluorescent lamp or a fluorescent tube is a low pressure mercury-vapor gas-discharge lamp that uses fluorescence to produce visible light. An electric current in the gas excites mercury vapor which produces short-wave ultraviolet light that then causes a phosphor coating on the inside of the lamp to glow. A fluorescent lamp converts electrical energy into useful light much more efficiently than incandescent lamps. The typical luminous efficacy of fluorescent lighting systems is 50–100 lumens per watt, several times the efficacy of incandescent bulbs with comparable light output, but less than that of a typical LED bulb.

Fluorescent lamps

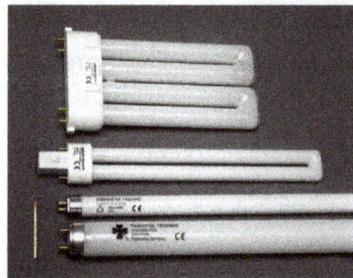

Top, two compact fluorescent lamps. Bottom, two fluorescent tube lamps. A matchstick, left, is shown for scale.

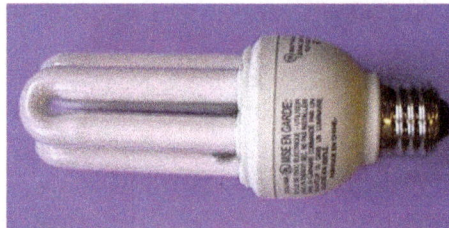

Compact fluorescent lamp with electronic ballast

Fluorescent lamp fixtures are more costly than incandescent lamps because they require a ballast to regulate the current through the lamp, but the lower energy cost typically offsets the higher initial cost. Compact fluorescent lamps are now available in the same popular sizes as incandescents and are used as an energy-saving alternative in homes.

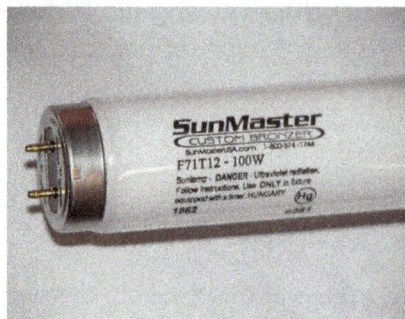

Typical F71T12 100 W bi-pin lamp used in tanning beds. The (Hg) symbol indicates that this lamp contains mercury. In the US, this symbol is now required on all fluorescent lamps that contain mercury.

Because they contain mercury, many fluorescent lamps are classified as hazardous waste. The United States Environmental Protection Agency recommends that fluorescent lamps be segregated from general waste for recycling or safe disposal, and some jurisdictions require recycling of them.

One style of lamp holder for T12 and T8 bi pin fluorescent lamps

Inside the lamp end of a preheat bi-pin lamp. In this lamp the filament is surrounded by an oblong metal cathode shield, which helps reduce lamp end darkening.

History

Physical Discoveries

Fluorescence of certain rocks and other substances had been observed for hundreds of years before its nature was understood. By the middle of the 19th century, experimenters had observed a radiant glow emanating from partially evacuated glass vessels through which an electric current passed. One of the first to explain it was the Irish scientist Sir George Stokes from the University of Cambridge, who named the phenomenon "fluorescence" after fluorite, a mineral many of whose samples glow strongly due to impurities. The explanation relied on the nature of electricity and light phenomena as developed by the British scientists Michael Faraday in the 1840s and James Clerk Maxwell in the 1860s.

Little more was done with this phenomenon until 1856 when a German glassblower named Heinrich Geissler created a mercury vacuum pump that evacuated a glass tube to an extent not previously possible. When an electric current passed through a Geissler tube, a strong green glow on the walls of the tube at the cathode end could be observed. Because it produced some beautiful light effects, the Geissler tube was a popular source of amusement. More important, however, was its contribution to scientific research. One of the first scientists to experiment with a Geissler tube was Julius Plücker who systematically described in 1858 the luminescent effects that occurred in a Geissler tube. He also made the important observation that the glow in the tube shifted position when in proximity to an electromagnetic field. Alexandre Edmond Becquerel observed in 1859 that certain substances gave off light when they were placed in a Geissler tube. He went on to apply thin coatings of luminescent materials to the surfaces of these tubes. Fluorescence occurred, but the tubes were very inefficient and had a short operating life.

Inquiries that began with the Geissler tube continued as even better vacuums were produced. The most famous was the evacuated tube used for scientific research by William Crookes. That tube was evacuated by the highly effective mercury vacuum pump created by Hermann Sprengel. Research conducted by Crookes and others ultimately led to the discovery of the electron in 1897 by J. J. Thomson and X-rays in 1895 by Wilhelm Roentgen. But the Crookes tube, as it came to be known, produced little light because the vacuum in it was too good and thus lacked the trace amounts of gas that are needed for electrically stimulated luminescence.

Early Discharge Lamps

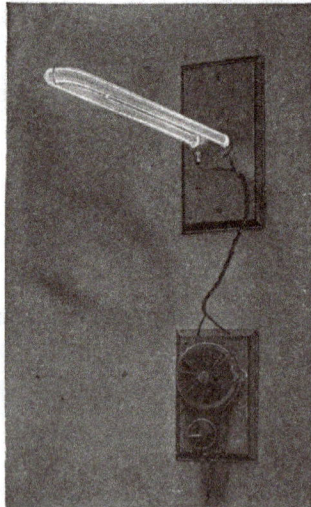

One of the first mercury vapor lamps invented by Peter Cooper Hewitt, 1903. It was similar to a fluorescent lamp without the fluorescent coating on the tube, and produced greenish light. The round device under the lamp is the ballast.

While Becquerel was interested primarily in conducting scientific research into fluorescence, Thomas Edison briefly pursued fluorescent lighting for its commercial potential. He invented a fluorescent lamp in 1896 that used a coating of calcium tungstate as the fluorescing substance, excited by X-rays, but although it received a patent in 1907, it was not put into production. As with a few other attempts to use Geissler tubes for illumination, it had a short operating life, and given the success of the incandescent light, Edison had little reason to pursue an alternative means of electrical illumination. Nikola Tesla made similar experiments in the 1890s, devising high-frequency powered fluorescent bulbs that gave a bright greenish light, but as with Edison's devices, no commercial success was achieved.

Although Edison lost interest in fluorescent lighting, one of his former employees was able to create a gas-based lamp that achieved a measure of commercial success. In 1895 Daniel McFarlan Moore demonstrated lamps 2 to 3 meters (6.6 to 9.8 ft) in length that used carbon dioxide or nitrogen to emit white or pink light, respectively. As with future fluorescent lamps, they were considerably more complicated than an incandescent bulb.

Peter Cooper Hewitt

After years of work, Moore was able to extend the operating life of the lamps by inventing an electromagnetically controlled valve that maintained a constant gas pressure within the tube. Although Moore's lamp was complicated, was expensive to install, and required very high voltages, it was considerably more efficient than incandescent lamps, and it produced a closer approximation to natural daylight than contemporary incandescent lamps. From 1904 onwards Moore's lighting system was installed in a number of stores and offices. Its success contributed to General Electric's motivation to improve the incandescent lamp, especially its filament. GE's efforts came to fruition with the invention of a tungsten-based filament. The extended lifespan and improved efficacy of incandescent bulbs negated one of the key advantages of Moore's lamp, but GE purchased the relevant patents in 1912. These patents and the inventive efforts that supported them were to be of considerable value when the firm took up fluorescent lighting more than two decades later.

At about the same time that Moore was developing his lighting system, another American was creating a means of illumination that also can be seen as a precursor to the modern fluorescent lamp. This was the mercury-vapor lamp, invented by Peter Cooper Hewitt and patented in 1901 (US 682692; this patent number is frequently misquoted as US 889,692). Hewitt's lamp glowed when an electric current was passed through mercury vapor at a low pressure. Unlike Moore's lamps, Hewitt's were manufactured in standardized sizes and operated at low voltages. The mercury-vapor lamp was superior to the incandescent lamps of the time in terms of energy efficiency, but the blue-green light it produced limited its applications. It was, however, used for photography and some industrial processes.

Mercury vapor lamps continued to be developed at a slow pace, especially in Europe, and by the early 1930s they received limited use for large-scale illumination. Some of them employed fluorescent coatings, but these were used primarily for color correction and not for enhanced light output. Mercury vapor lamps also anticipated the fluorescent lamp in their incorporation of a ballast to maintain a constant current.

Cooper-Hewitt had not been the first to use mercury vapor for illumination, as earlier efforts had been mounted by Way, Rapieff, Arons, and Bastian and Salisbury. Of particular importance was the mercury vapor lamp invented by Küch in Germany. This lamp used quartz in place of glass to allow higher operating temperatures, and hence greater efficiency. Although its light output relative to electrical consumption was better than that of other sources of light, the light it produced was similar to that of the Cooper-Hewitt lamp in that it lacked the red portion of the spectrum, making it unsuitable for ordinary lighting.

Neon Lamps

The next step in gas-based lighting took advantage of the luminescent qualities of neon, an inert gas that had been discovered in 1898 by isolation from the atmosphere. Neon glowed a brilliant red when used in Geissler tubes. By 1910, Georges Claude, a Frenchman who had developed a technology and a successful business for air liquefaction, was obtaining enough neon as a byproduct to support a neon lighting industry. While neon lighting was used around 1930 in France for general illumination, it was no more energy-efficient than conventional incandescent lighting. Neon tube lighting, which also includes the use of argon and mercury vapor as alternate gases, came to be used primarily for eye-catching signs and advertisements. Neon lighting was relevant to the development of fluorescent lighting, however, as Claude's improved electrode (patented in 1915) overcame "sputtering", a major source of electrode degradation. Sputtering occurred when ionized particles struck an electrode and tore off bits of metal. Although Claude's invention required electrodes with a lot of surface area, it showed that a major impediment to gas-based lighting could be overcome.

The development of the neon light also was significant for the last key element of the fluorescent lamp, its fluorescent coating. In 1926 Jacques Risler received a French patent for the application of fluorescent coatings to neon light tubes. The main use of these lamps, which can be considered the first commercially successful fluorescents, was for advertising, not general illumination. This, however, was not the first use of fluorescent coatings; Becquerel had earlier used the idea and Edison used calcium tungstate for his unsuccessful lamp. Other efforts had been mounted, but all were plagued by low efficiency and various technical problems. Of particular importance was the invention in 1927 of a low-voltage "metal vapor lamp" by Friedrich Meyer, Hans-Joachim Spanner, and Edmund Germer, who were employees of a German firm in Berlin. A German patent was granted but the lamp never went into commercial production.

Commercialization of Fluorescent Lamps

All the major features of fluorescent lighting were in place at the end of the 1920s. Decades of invention and development had provided the key components of fluorescent lamps: economically manufactured glass tubing, inert gases for filling the tubes, electrical ballasts, long-lasting electrodes, mercury vapor as a source of luminescence, effective means of producing a reliable electrical discharge, and fluorescent coatings that could be energized by ultraviolet light. At this point, intensive development was more important than basic research.

In 1934, Arthur Compton, a renowned physicist and GE consultant, reported to the GE lamp department on successful experiments with fluorescent lighting at General Electric Co., Ltd. in Great Britain (unrelated to General Electric in the United States). Stimulated by this report, and with all of the key elements available, a team led by George E. Inman built a prototype fluorescent lamp in 1934 at General Electric's Nela Park (Ohio) engineering laboratory. This was not a trivial exercise; as noted by Arthur A. Bright, "A great deal of experimentation had to be done on lamp sizes and shapes, cathode construction, gas pressures of both argon and mercury vapor, colors of fluorescent powders, methods of attaching them to the inside of the tube, and other details of the lamp and its auxiliaries before the new device was ready for the public."

In addition to having engineers and technicians along with facilities for R&D work on fluorescent lamps, General Electric controlled what it regarded as the key patents covering fluorescent lighting, including the patents originally issued to Hewitt, Moore, and Küch. More important than these was a patent covering an electrode that did not disintegrate at the gas pressures that ultimately were employed in fluorescent lamps. Albert W. Hull of GE's Schenectady Research Laboratory filed for a patent on this invention in 1927, which was issued in 1931. General Electric used its control of the patents to prevent competition with its incandescent lights and probably delayed the introduction of fluorescent lighting by 20 years. Eventually, war production required 24-hour factories with economical lighting and fluorescent lights became available.

While the Hull patent gave GE a basis for claiming legal rights over the fluorescent lamp, a few months after the lamp went into production the firm learned of a U.S. patent application that had been filed in 1927 for the aforementioned "metal vapor lamp" invented in Germany by Meyer, Spanner, and Germer. The patent application indicated that the lamp had been created as a superior means of producing ultraviolet light, but the application also contained a few statements referring to fluorescent illumination. Efforts to obtain a U.S. patent had met with numerous delays, but were it to be granted, the patent might have caused serious difficulties for GE. At first, GE sought to block the issuance of a patent by claiming that priority should go to one of their employees, Leroy J. Buttolph, who according to their claim had invented a fluorescent lamp in 1919 and whose patent application was still pending. GE also had filed a patent application in 1936 in Inman's name to cover the "improvements" wrought by his group. In 1939 GE decided that the claim of Meyer, Spanner, and Germer had some merit, and that in any event a long interference procedure was not in their best interest. They therefore dropped the Buttolph claim and paid $180,000 to acquire the Meyer, et al. application, which at that point was owned by a firm known as Electrons, Inc. The patent was duly awarded in December 1939. This patent, along with the Hull patent, put GE on what seemed to be firm legal ground, although it faced years of legal challenges from Sylvania Electric Products, Inc., which claimed infringement on patents that it held.

Even though the patent issue would not be completely resolved for many years, General Electric's strength in manufacturing and marketing gave it a pre-eminent position in the emerging fluorescent light market. Sales of "fluorescent lumiline lamps" commenced in 1938 when four different sizes of tubes were put on the market. They were used in fixtures manufactured by three leading corporations, Lightolier, Artcraft Fluorescent Lighting Corporation, and Globe Lighting. The Slimline fluorescent ballast's public introduction in 1946 was by Westinghouse and General Electric and Showcase/Display Case fixtures were introduced by Artcraft Fluorescent Lighting Corporation in 1946. During the following year, GE and Westinghouse publicized the new lights through exhibitions at the New York World's Fair and the Golden Gate International Exposition in San Francisco. Fluorescent lighting systems spread rapidly during World War II as wartime manufacturing intensified lighting demand. By 1951 more light was produced in the United States by fluorescent lamps than by incandescent lamps.

In the first years zinc orthosilicate with varying content of beryllium was used as greenish phosphor. Small additions of magnesium tungstate improved the blue part of the spectrum yielding acceptable white. After it was discovered that beryllium was toxic, halophosphate based phosphors took over.

Principles of Operation

The fundamental means for conversion of electrical energy into radiant energy in a fluorescent lamp relies on inelastic scattering of electrons when an incident electron collides with an atom in the gas. If the (incident) free electron has enough kinetic energy, it transfers energy to the atom's outer electron, causing that electron to temporarily jump up to a higher energy level. The collision is 'inelastic' because a loss of kinetic energy occurs.

This higher energy state is unstable, and the atom will emit an ultraviolet photon as the atom's electron reverts to a lower, more stable, energy level. Most of the photons that are released from the mercury atoms have wavelengths in the ultraviolet (UV) region of the spectrum, predominantly at wavelengths of 253.7 and 185 nanometers (nm). These are not visible to the human eye, so they must be converted into visible light. This is done by making use of fluorescence. Ultraviolet photons are absorbed by electrons in the atoms of the lamp's interior fluorescent coating, causing a similar energy jump, then drop, with emission of a further photon. The photon that is emitted from this second interaction has a lower energy than the one that caused it. The chemicals that make up the phosphor are chosen so that these emitted photons are at wavelengths visible to the human eye. The difference in energy between the absorbed ultra-violet photon and the emitted visible light photon goes toward heating up the phosphor coating.

When the light is turned on, the electric power heats up the cathode enough for it to emit electrons (thermionic emission). These electrons collide with and ionize noble gas atoms inside the bulb surrounding the filament to form a plasma by the process of impact ionization. As a result of avalanche ionization, the conductivity of the ionized gas rapidly rises, allowing higher currents to flow through the lamp.

The fill gas helps determine the operating electrical characteristics of the lamp, but does not give off light itself. The fill gas effectively increases the distance that electrons travel through the tube, which allows an electron a greater chance of interacting with a mercury atom. Argon atoms, excited to a metastable state by impact of an electron, can impart this energy to a neutral mercury atom and ionize it, described as the Penning effect. This has the benefit of lowering the breakdown and operating voltage of the lamp, compared to other possible fill gases such as krypton.

Construction

A fluorescent lamp tube is filled with a gas containing low pressure mercury vapor and argon, xenon, neon, or krypton. The pressure inside the lamp is around 0.3% of atmospheric pressure. The inner surface of the lamp is coated with a fluorescent (and often slightly phosphorescent) coating made of varying blends of metallic and rare-earth phosphor salts. The lamp's electrodes are typically made of coiled tungsten and usually referred to as cathodes because of their prime function of emitting electrons. For this, they are coated with a mixture of barium, strontium and calcium oxides chosen to have a low thermionic emission temperature.

Close-up of the cathodes of a germicidal lamp (an essentially similar design that uses no fluorescent phosphor, allowing the electrodes to be seen.)

Fluorescent lamp tubes are typically straight and range in length from about 100 millimeters (3.9 in) for miniature lamps, to 2.43 meters (8.0 ft) for high-output lamps. Some lamps have the tube bent into a circle, used for table lamps or other places where a more compact light source is desired. Larger U-shaped lamps are used to provide the same amount of light in a more compact area, and are used for special architectural purposes. Compact fluorescent lamps have several small-diameter tubes joined in a bundle of two, four, or six, or a small diameter tube coiled into a helix, to provide a high amount of light output in little volume.

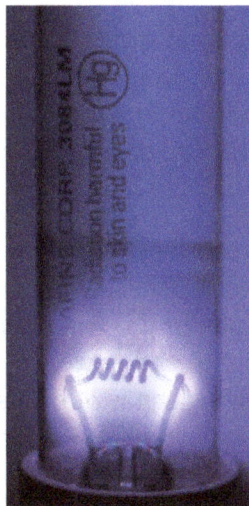

A germicidal lamp uses a low pressure mercury vapor glow discharge identical to that in a fluorescent lamp, but the germicidal lamp uses an uncoated fused quartz envelope so the ultraviolet radiation can escape.

Light-emitting phosphors are applied as a paint-like coating to the inside of the tube. The organic solvents are allowed to evaporate, then the tube is heated to nearly the melting point of glass to drive off remaining organic compounds and fuse the coating to the lamp tube. Careful control of the grain size of the suspended phosphors is necessary; large grains, 35 micrometers or larger, lead to weak grainy coatings, whereas too many small particles 1 or 2 micrometers or smaller leads to poor light maintenance and efficiency. Most phosphors perform best with a particle size around 10 micrometers. The coating must be thick enough to capture all the ultraviolet light produced by the mercury arc, but not so thick that the phosphor coating absorbs too much visible light. The first phosphors were synthetic versions of naturally occurring fluorescent minerals, with small amounts of metals added as activators. Later other compounds were discovered, allowing differing colors of lamps to be made.

Electrical Aspects of Operation

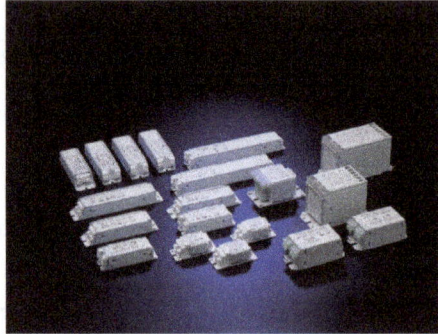

Different ballasts for fluorescent and discharge lamps

Fluorescent lamps are negative differential resistance devices, so as more current flows through them, the electrical resistance of the fluorescent lamp drops, allowing for even more current to flow. Connected directly to a constant-voltage power supply, a fluorescent lamp would rapidly self-destruct due to the uncontrolled current flow. To prevent this, fluorescent lamps must use an auxiliary device, a ballast, to regulate the current flow through the lamp.

The terminal voltage across an operating lamp varies depending on the arc current, tube diameter, temperature, and fill gas. A fixed part of the voltage drop is due to the electrodes. A general lighting service 48-inch (1,219 mm) T12 lamp operates at 430 mA, with 100 volts drop. High output lamps operate at 800 mA, and some types operate up to 1.5 A. The power level varies from 33 to 82 watts per meter of tube length (10 to 25 W/ft) for T12 lamps.

The simplest ballast for alternating current (AC) use is an inductor placed in series, consisting of a winding on a laminated magnetic core. The inductance of this winding limits the flow of AC current. This type is still used, for example, in 120 volt operated desk lamps using relatively short lamps. Ballasts are rated for the size of lamp and power frequency. Where the AC voltage is insufficient to start long fluorescent lamps, the ballast is often a step-up autotransformer with substantial leakage inductance (so as to limit the current flow). Either form of inductive ballast may also include a capacitor for power factor correction.

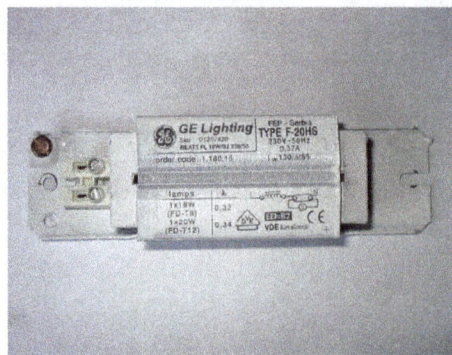

230 V ballast for 18–20 W

Many different circuits have been used to operate fluorescent lamps. The choice of circuit is based on AC voltage, tube length, initial cost, long term cost, instant versus non-instant starting, temperature ranges and parts availability, etc.

Fluorescent lamps can run directly from a direct current (DC) supply of sufficient voltage to strike an arc. The ballast must be resistive, and would consume about as much power as the lamp. When operated from DC, the starting switch is often arranged to reverse the polarity of the supply to the lamp each time it is started; otherwise, the mercury accumulates at one end of the tube. Fluorescent lamps are (almost) never operated directly from DC for those reasons. Instead, an inverter converts the DC into AC and provides the current-limiting function as described below for electronic ballasts.

Effect of Temperature

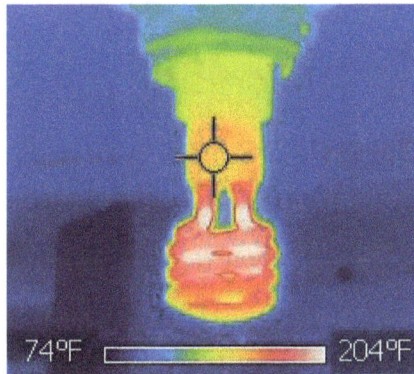

Thermal image of a helical fluorescent lamp.

The light output and performance of fluorescent lamps is critically affected by the temperature of the bulb wall and its effect on the partial pressure of mercury vapor within the lamp. Each lamp contains a small amount of mercury, which must vaporize to support the lamp current and generate light. At low temperatures the mercury is in the form of dispersed liquid droplets. As the lamp warms, more of the mercury is in vapor form. At higher temperatures, self-absorption in the vapor reduces the yield of UV and visible light. Since mercury condenses at the coolest spot in the lamp, careful design is required to maintain that spot at the optimum temperature, around 40 °C (104 °F).

By using an amalgam with some other metal, the vapor pressure is reduced and the optimum temperature range extended upward; however, the bulb wall "cold spot" temperature must still be controlled to prevent migration of the mercury out of the amalgam and condensing on the cold spot. Fluorescent lamps intended for higher output will have structural features such as a deformed tube or internal heat-sinks to control cold spot temperature and mercury distribution. Heavily loaded small lamps, such as compact fluorescent lamps, also include heat-sink areas in the tube to maintain mercury vapor pressure at the optimum value.

Losses

Only a fraction of the electrical energy input into a lamp is converted to useful light. The ballast dissipates some heat; electronic ballasts may be around 90% efficient. A fixed voltage drop occurs at the electrodes, which also produces heat. Some of the energy in the mercury vapor column is also dissipated, but about 85% is turned into visible and ultraviolet light.

40 watts power in

Ballast 90%   4 w Ballast loss

Electrodes 92%   3 w Electrode loss

Discharge 85%   5 w Not visible or UV

Phosphor 86%   4 w UV photons lost

Quantum efficiency 45%   13 w Quantum efficiency
                                               5.5 ev UV to 2.5 ev visible

11 w

Visible light   Losses in a 36 w t8 tri-phosphor fluorescent lamp
out   with electronic ballast

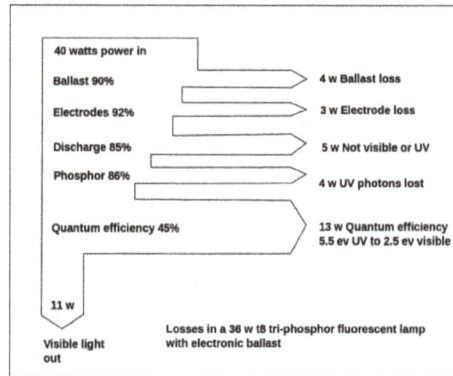

A Sankey diagram of energy losses in a fluorescent lamp. In modern designs, the biggest loss is the quantum efficiency of converting high-energy UV photons to lower-energy visible light photons.

The UV light is absorbed by the lamp's fluorescent coating, which re-radiates the energy at longer wavelengths to emit visible light. Not all the UV energy striking the phosphor gets converted into visible light. In a modern lamp, for every 100 incident photons of UV impacting the phosphor, only 86 visible light photons are emitted (a quantum efficiency of 86%). The largest single loss in modern lamps is due to the lower energy of each photon of visible light, compared to the energy of the UV photons that generated them (a phenomenon called Stokes shift). Incident photons have an energy of 5.5 electron volts but produce visible light photons with energy around 2.5 electron volts, so only 45% of the UV energy is used; the rest is dissipated as heat. If a so-called "two-photon" phosphor could be developed, this would improve the efficiency but much research has not yet found such a system.

Cold-Cathode Fluorescent Lamps

Most fluorescent lamps use electrodes that operate by thermionic emission, meaning they are operated at a high enough temperature for the electrode material (usually aided by a special coating) to emit electrons into the tube by heat.

However, there are also tubes that operate in cold cathode mode, whereby electrons are liberated into the tube only by the large potential difference (voltage) between the electrodes. This does not mean the electrodes are cold (indeed, they can be very hot), but it does mean they are operating below their thermionic emission temperature. Because cold cathode lamps have no thermionic emission coating to wear out they can have much longer lives than hot cathode tubes. This quality makes them desirable for maintenance-free long-life applications (such as backlights in liquid crystal displays). Sputtering of the electrode may still occur, but electrodes can be shaped (e.g. into an internal cylinder) to capture most of the sputtered material so it is not lost from the electrode.

Cold cathode lamps are generally less efficient than thermionic emission lamps because the cathode fall voltage is much higher. The increased fall voltage results in more power dissipation at tube ends, which does not contribute to light output. However, this is less significant with longer tubes. The increased power dissipation at tube ends also usually means cold cathode tubes have to be run at a lower loading than their thermionic emission equivalents. Given the higher tube voltage required anyway, these tubes can easily be made long, and even run as series strings. They are better suited for bending into special shapes for lettering and signage, and can also be instantly switched on or off.

Starting

The noble gas used in the fluorescent tube (commonly argon) must be ionized before the arc can "strike" within the tube. For small lamps, it does not take much voltage to strike the arc and starting the lamp presents no problem, but larger tubes require a substantial voltage (in the range of a thousand volts).

Preheating

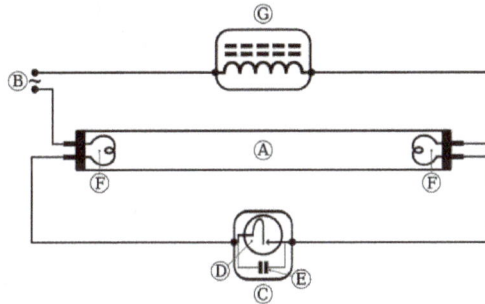

A *preheat* fluorescent lamp circuit using an automatic starting switch. A: Fluorescent tube, B: Power (+220 volts), C: Starter, D: Switch (bi-metallic thermostat), E: Capacitor, F: Filaments, G: Ballast

Starting a preheat lamp. The automatic starter switch flashes orange each time it attempts to start the lamp.

This technique uses a combination filament–cathode at each end of the lamp in conjunction with a mechanical or automatic (bi-metallic) switch that initially connect the filaments in series with the ballast to preheat them; when the arc is struck the filaments are disconnected. This system is described as *preheat* in some countries and *switchstart* in others. These systems are standard equipment in 200–240 V countries (and for 100–120 V lamps up to about 30 watts).

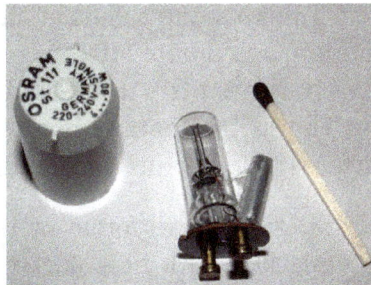

A *preheat* fluorescent lamp "starter" (automatic starting switch)

Before the 1960s four-pin thermal starters and manual switches were used. A mechanism then widely used for preheating, still in common use, is a glow switch starter (illustrated). It consists of a normally open bi-metallic switch in a small sealed gas-discharge lamp containing inert gas (neon or argon).

Electronic fluorescent lamp starters

When power is first applied to the circuit, there will be a glow discharge across the electrodes in the starter lamp. This heats the gas in the starter and causes one of the bi-metallic contacts to bend towards the other. When the contacts touch, the two filaments of the fluorescent lamp and the ballast will effectively be switched in series to the supply voltage. The current through the filaments causes them to heat up and emit electrons into the tube gas by thermionic emission. In the starter, the touching contacts short out the voltage sustaining the glow discharge, extinguishing it so the gas cools down and no longer heats the bi-metallic switch, which opens within a second or two. The current through the filaments and the inductive ballast is abruptly interrupted, leaving the full line voltage applied between the filaments at the ends of the tube and generating an inductive kick which provides the high voltage needed to start the lamp. The lamp will fail to strike if the filaments are not hot enough, in which case the cycle repeats; several cycles are usually needed, which causes flickering and clicking during starting (older thermal starters behaved better in this respect). A power factor correction (PFC) capacitor draws leading current from the mains to compensate for the lagging current drawn by the lamp circuit.

Once the tube strikes, the impinging main discharge keeps the cathodes hot, permitting continued electron emission without the need for the filaments to continue to be heated. The starter switch does not close again because the voltage across the lit tube is insufficient to start a glow discharge in the starter.

With automated starters such as glow starters, a failing tube will cycle endlessly, flickering as the lamp quickly goes out because the emission mix is insufficient to keep the lamp current high enough to keep the glow starter open. This runs the ballast at higher temperature. Some more advanced starters time out in this situation, and do not attempt repeated starts until power is reset. Some older systems used a thermal over-current trip to detect repeated starting attempts and disable the circuit until manually reset. The switch contacts in glow starters are subject to wear and inevitably fail eventually, so the starter is manufactured as a plug-in replaceable unit.

More recently introduced electronic starters use a different method to preheat the cathodes. They may be designed to be plug-in interchangeable with glow starters for use in standard fittings. They commonly use a purpose-designed semiconductor switch and "soft start" the lamp by preheating the cathodes before applying a controlled starting pulse which strikes the lamp first time without flickering; this dislodges a minimal amount of material from the cathodes during starting, giving longer lamp life than possible with the uncontrolled impulses to which the lamp is subjected in a switchstart. This is claimed to prolong lamp life by a factor of typically 3 to 4 times for a lamp frequently switched on as in domestic use, and to reduce the blackening of the ends of the lamp typical of fluorescent tubes. The circuit is typically complex, but the complexity is built into the IC. Electronic starters may be optimized for fast starting (typical start time of 0.3 seconds), or for most reliable starting even at low temperatures and with low supply voltages, with a startup time of 2–4 seconds. The faster-start units may produce audible noise during start-up.

Electronic starters only attempt to start a lamp for a short time when power is initially applied, and do not repeatedly attempt to restrike a lamp that is dead and unable to sustain an arc; some automatically shut down a failed lamp. This eliminates the re-striking of a lamp and the continuous flickering of a failing lamp with a glow starter. Electronic starters are not subject to wear and do not need replacing periodically, although they may fail like any other electronic circuit. Manufacturers typically quote lives of 20 years, or as long as the light fitting. Starters are inexpensive, typically less than 50 US cents for the short-lived glow type (depending upon lamp power), and perhaps ten times more for the electronic type as of 2013.

Instant Start

Another type of tube does not have filaments to start it at all. *Instant start* fluorescent tubes simply use a high enough voltage to break down the gas and mercury column and thereby start arc conduction. These tubes can be identified by a single pin at each end of the tube. The lamp holders have a "disconnect" socket at the low-voltage end which disconnects the ballast when the tube is removed, to prevent electric shock. Low-cost lighting fixtures with an integrated electronic ballast use instant start on lamps originally designed for preheating, although it shortens lamp life.

Rapid Start

Newer *rapid start* ballast designs provide filament power windings within the ballast; these rapidly and continuously warm the filaments/cathodes using low-voltage AC. Usually operating at a lower arc voltage than the instant start design; no inductive voltage spike is produced for starting, so the lamps must be mounted near a grounded (earthed) reflector to allow the glow discharge to propagate through the tube and initiate the arc discharge. In some lamps a grounded "starting aid" strip is attached to the outside of the lamp glass.

A rapid-start "iron" (magnetic) ballast continually heats the cathodes at the ends of the lamps. This ballast runs two F40T12 lamps in series.

Quick-Start

Quick-start ballasts use a small auto-transformer to heat the filaments when power is first applied. When an arc strikes, the filament heating power is reduced and the tube will start within half a second. The auto-transformer is either combined with the ballast or may be a separate unit. Tubes need to be mounted near an earthed metal reflector in order for them to strike. Quick-start ballasts are more common in commercial installations because of lower maintenance costs. A quick-start ballast eliminates the need for a starter switch, a common source of lamp failures. Nonetheless, Quick-start ballasts are also used in domestic (residential) installations because of the desirable feature that a Quick-start ballast light turns on nearly immediately after power is applied (when a switch is turned on). Quick-start ballasts are used only on 240 V circuits and are designed for use with the older, less efficient T12 tubes.

Semi-Resonant Start

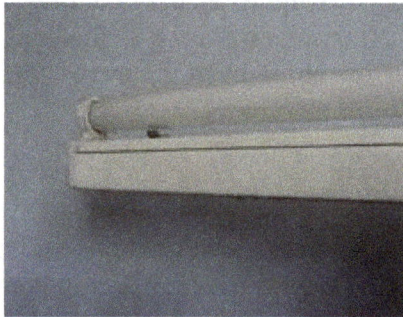

A 65-watt fluorescent lamp starting on a semi-resonant start circuit

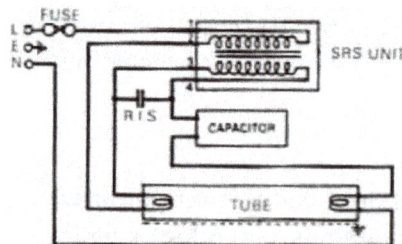

Semi-resonant start circuit.

A semi-resonant start circuit diagram

The semi-resonant start circuit was invented by Thorn Lighting for use with T12 fluorescent tubes. This method uses a double wound transformer and a capacitor. With no arc current, the transformer and capacitor resonate at line frequency and generate about twice the supply voltage across the tube, and a small electrode heating current. This tube voltage is too low to strike the arc with cold electrodes, but as the electrodes heat up to thermionic emission temperature, the tube striking voltage falls below that of the ringing voltage, and the arc strikes. As the electrodes heat, the lamp slowly, over three to five seconds, reaches full brightness. As the arc current increases and tube voltage drops, the circuit provides current limiting.

Semi-resonant start circuits are mainly restricted to use in commercial installations because of the higher initial cost of circuit components. However, there are no starter switches to be replaced and cathode damage is reduced during starting making lamps last longer, reducing maintenance costs. Due to the high open circuit tube voltage, this starting method is particularly good for starting tubes in cold locations. Additionally, the circuit power factor is almost 1.0, and no additional power factor correction is needed in the lighting installation. As the design requires that twice the supply voltage must be lower than the cold-cathode striking voltage (or the tubes would erroneously instant-start), this design cannot be used with 240 volt AC power unless the tubes are at least 1.2 m (3 ft 11 in) length. Semi-resonant start fixtures are generally incompatible with energy saving T8 retrofit tubes, because such tubes have a higher starting voltage than T12 lamps and may not start reliably, especially in low temperatures. Recent proposals in some countries to phase out T12 tubes will reduce the application of this starting method.

Programmed Start

This is used with electronic ballasts shown below. A programmed-start ballast is a more advanced version of rapid start. This ballast applies power to the filaments first, then after a short delay to allow the cathodes to preheat, applies voltage to the lamps to strike an arc. This ballast gives the best life and most starts from lamps, and so is preferred for applications with very frequent power cycling such as vision examination rooms and restrooms with a motion detector switch.

Electronic Ballasts

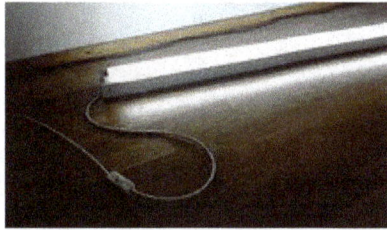

Fluorescent lamp with an electronic ballast.

Electronic ballasts employ transistors to change the supply frequency into high-frequency AC while also regulating the current flow in the lamp. Some still use an inductance to limit the current, but the higher frequency allows a much smaller inductance to be used. Others use a capacitor-transistor combination to replace the inductor, since a transistor and capacitor working together can simulate the action of an inductor. These ballasts take advantage of the higher efficacy of lamps operated with higher-frequency current, which rises by almost 10% at 10 kHz, compared to efficacy at normal power frequency. When the AC period is shorter than the relaxation time to de-ionize mercury atoms in the discharge column, the discharge stays closer to optimum operating condition. Electronic ballasts typically work in rapid start or instant start mode. Electronic ballasts are commonly supplied with AC power, which is internally converted to DC and then back to a variable frequency AC waveform. Depending upon the capacitance and the quality of constant-current pulse-width modulation, this can largely eliminate modulation at 100 or 120 Hz.

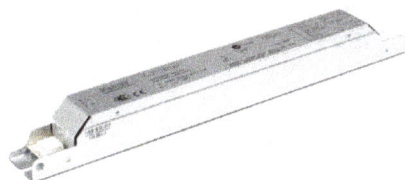

Electronic ballast for fluorescent lamp, 2×58 W

Low cost ballasts mostly contain only a simple oscillator and series resonant LC circuit. When turned on, the oscillator starts, and resonant current excites the LC circuit. This resonant current directly drives a switching transistor through a ring core transformer. This principle is called the current resonant inverter circuit. After a short time the voltage across the lamp reaches about 1 kV and the lamp ignites. The process is too fast to preheat the cathodes, so the lamp instant-starts in cold cathode mode. The cathode filaments are still used for protection of the ballast from overheating if the lamp does not ignite. A few manufacturers use positive temperature coefficient (PTC) thermistors to disable instant starting and give some time to preheat the filaments.

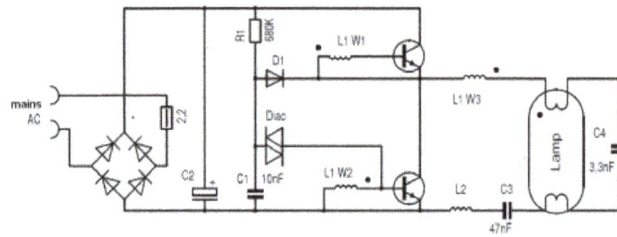

Electronic ballast basic schematic

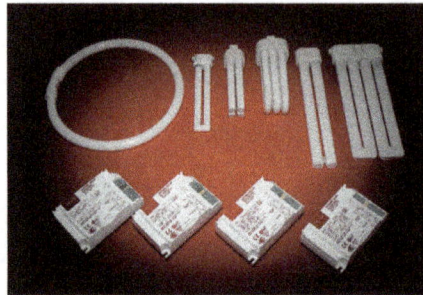

Electronic ballasts and different compact fluorescent lamps

More complex electronic ballasts use programmed start. The output frequency is started above the resonance frequency of the output circuit of the ballast; and after the filaments are heated, the frequency is rapidly decreased. If the frequency approaches the resonant frequency of the ballast, the output voltage will increase so much that the lamp will ignite. If the lamp does not ignite, an electronic circuit stops the operation of the ballast.

Many electronic ballasts are controlled by a microcontroller or similar, and these are sometimes called digital ballasts. Digital ballasts can apply quite complex logic to lamp starting and operation. This enables functions such as testing for broken electrodes and missing tubes before attempting to start, auto detect tube replacement, and auto detection of tube type, such that a single ballast can be used with several different tubes, even those that operate at different arc currents, etc. Once such fine grained control over the starting and arc current is achievable, features such as dimming, and having the ballast maintain a constant light level against changing sunlight contribution are all easily included in the embedded microcontroller software, and can be found in various manufacturers' products.

Since introduction in the 1990s, high-frequency ballasts have been used in general lighting fixtures with either rapid start or pre-heat lamps. These ballasts convert the incoming power to an output frequency in excess of 20 kHz. This increases lamp efficiency. These are used in several applications, including new generation tanning lamp systems, whereby a 100 watt lamp (e.g., F71T12BP) can be lit using 90 watts of actual power while obtaining the same luminous flux (measured in lumens) as magnetic ballasts. These ballasts operate with voltages that can be almost 600 volts, requiring some consideration in housing design, and can cause a minor limitation in the length of the wire leads from the ballast to the lamp ends.

End of Life

The end of life failure mode for fluorescent lamps varies depending on how they are used and their control gear type. Often the light will turn pink with black burns on the ends of the lamp due to sputtering of emission mix. The lamp may also flicker at a noticeable rate. More information about normal failure modes are as follows:

Emission Mix

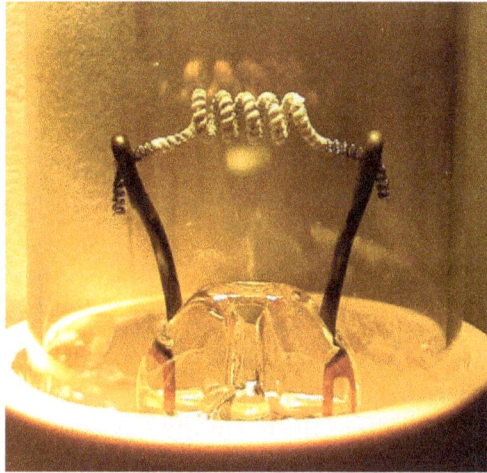

Closeup of the filament on a low pressure mercury gas discharge lamp showing white thermionic emission mix coating on the central portion of the coil acting as hot cathode. Typically made of a mixture of barium, strontium and calcium oxides, the coating is sputtered away through normal use, often eventually resulting in lamp failure.

The "emission mix" on the lamp filaments/cathodes is required to enable electrons to pass into the gas via thermionic emission at the lamp operating voltages used. The mix is slowly sputtered off by bombardment with electrons and mercury ions during operation, but a larger amount is sputtered off each time the lamp is started with cold cathodes. The method of starting the lamp has a significant impact on this. Lamps operated for typically less than 3 hours each switch-on will normally run out of the emission mix before other parts of the lamp fail. The sputtered emission mix forms the dark marks at the lamp ends seen in old lamps. When all the emission mix is gone, the cathode cannot pass sufficient electrons into the gas fill to maintain the gas discharge at the designed lamp operating voltage. Ideally, the control gear should shut down the lamp when this happens. However, due to cost, negative differential resistance and sometimes high starting voltage, some control gear will provide sufficient increased operating voltage to continue lighting the lamp in cold cathode mode. This will cause overheating of the lamp end and rapid disintegration of the electrodes (filament goes open-circuit) and filament support wires until they are completely gone or the glass cracks, wrecking the low pressure gas fill and stopping the gas discharge.

Ballast Electronics

This may occur in compact fluorescent lamps with integral electrical ballasts or in linear lamps. Ballast electronics failure is a somewhat random process that follows the standard failure profile for any electronic device. There is an initial small peak of early failures, followed by a drop and steady increase over lamp life. Life of electronics is heavily dependent on operating temperature—it typically halves for each 10 °C temperature rise. The quoted average life of a lamp is usually at 25 °C (77 °F) ambient (this may vary by country). The average life of the electronics at this temperature is normally greater than this, so at this temperature, not many lamps will fail due to failure of the electronics. In some fittings, the ambient temperature could be well above this, in which case failure of the electronics may become the predominant failure mechanism. Similarly, running a compact fluorescent lamp base-up will result in hotter electronics, which can cause shorter average life (particularly with higher power rated ones). Electronic ballasts should be designed to shut down the tube when the emission mix runs out as described above. In the case of integral electronic ballasts, since they never have to work again, this is sometimes done by having them deliberately burn out some component to permanently cease operation.

In most CFLs the filaments are connected in series, with a small capacitor between them. The discharge, once lit, is in parallel to the capacitor and presents a lower-resistance path, effectively shorting the capacitor out.

Phosphor

The phosphor drops off in efficiency during use. By around 25,000 operating hours, it will typically be half the brightness of a new lamp (although some manufacturers claim much longer half-lives for their lamps). Lamps that do not suffer failures of the emission mix or integral ballast electronics will eventually develop this failure mode. They still work, but have become dim and inefficient. The process is slow, and often becomes obvious only when a new lamp is operating next to an old one.

Loss of Mercury

As in all mercury-based gas-filled tubes, mercury is slowly adsorbed into the glass, phosphor, and tube electrodes throughout the life of the lamp, until it can no longer function. Loss of mercury will take over from failure of the phosphor in some lamps. The failure symptoms are similar, except loss of mercury initially causes an extended run-up time to full light output, and finally causes the lamp to glow a dim pink when the mercury runs out and the argon base gas takes over as the primary discharge.

Subjecting the tube to asymmetric waveforms, where the total current flow through the tube does not cancel out and the tube effectively operates under a DC bias, causes asymmetric distribution of mercury ions along the tube due to cataphoresis. The localized depletion of mercury vapor pressure manifests as pink luminescence of the base gas in the vicinity of one of the electrodes, and the operating lifetime of the lamp may be dramatically shortened. This can be an issue with some poorly designed inverters.

Burned Filaments

The filaments can burn out (fail) at the end of the lamp's lifetime, opening the circuit and losing the capability to heat up. Both filaments lose function as they are connected in series, with just a simple switch start circuit a broken filament will render the lamp completely useless. Filaments rarely burn or fail open circuit unless the filament becomes depleted of emitter and the control gear is able to supply a high enough voltage across the tube to operate it in cold cathode mode. Some digital electronic ballasts are capable of detecting broken filaments and can still strike an arc with one or both filaments broken providing there is still sufficient emitter. A broken filament in a lamp attached to a magnetic ballast often causes both lamps to burn out or flicker.

Phosphors and The Spectrum of Emitted light

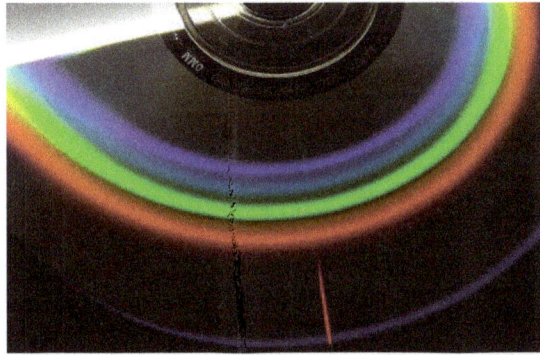

Light from a fluorescent tube lamp reflected by a CD shows the individual bands of color.

The spectrum of light emitted from a fluorescent lamp is the combination of light directly emitted by the mercury vapor, and light emitted by the phosphorescent coating. The spectral lines from the mercury emission and the phosphorescence effect give a combined spectral distribution of light that is different from those produced by incandescent sources. The relative intensity of light emitted in each narrow band of wavelengths over the visible spectrum is in different proportions compared to that of an incandescent source. Colored objects are perceived differently under light sources with differing spectral distributions. For example, some people find the color rendition produced by some fluorescent lamps to be harsh and displeasing. A healthy person can sometimes appear to have an unhealthy skin tone under fluorescent lighting. The extent to which this phenomenon occurs is related to the light's spectral composition, and may be gauged by its color rendering index (CRI).

Color Temperature

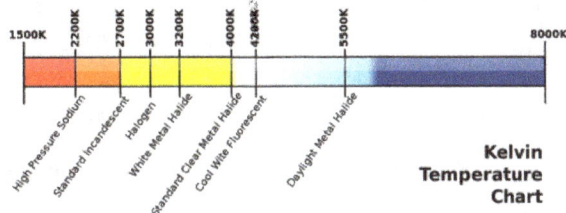

The color temperature of different electric lamps

Correlated color temperature (CCT) is a measure of the "shade" of whiteness of a light source compared with a blackbody. Typical incandescent lighting is 2700 K, which is yellowish-white. Halogen lighting is 3000 K. Fluorescent lamps are manufactured to a chosen CCT by altering the mixture of phosphors inside the tube. Warm-white fluorescents have CCT of 2700 K and are popular for residential lighting. Neutral-white fluorescents have a CCT of 3000 K or 3500 K. Cool-white fluorescents have a CCT of 4100 K and are popular for office lighting. Daylight fluorescents have a CCT of 5000 K to 6500 K, which is bluish-white.

High CCT lighting generally requires higher light levels. At dimmer illumination levels, the human eye perceives lower color temperatures as more pleasant, as related through the Kruithof curve. So, a dim 2700 K incandescent lamp appears comfortable and a bright 5000 K lamp also appears natural, but a dim 5000 K fluorescent lamp appears too pale. Daylight-type fluorescents look natural only if they are very bright.

Color Rendering Index

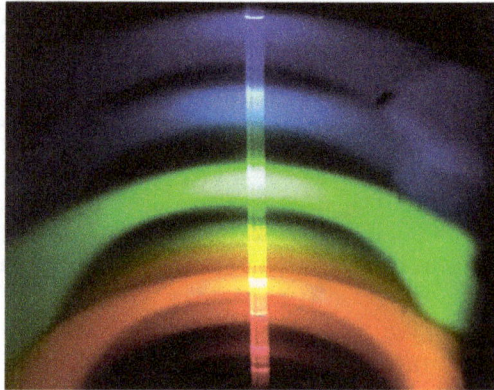

A helical cool-white fluorescent lamp reflected in a diffraction grating reveals the various spectral lines which make up the light.

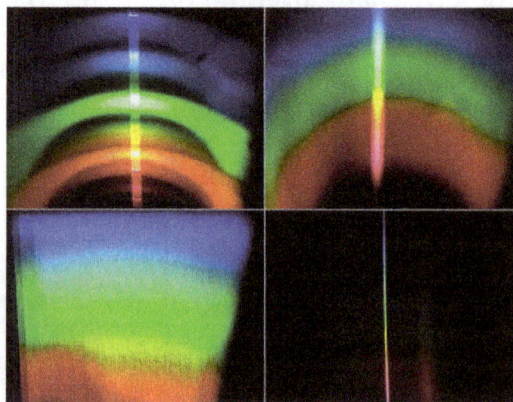

Fluorescent spectra in comparison with other forms of lighting. Clockwise from upper left: Fluorescent lamp, incandescent bulb, candle flame and LED lighting.

Color rendering index (CRI) is a measure of how well colors can be perceived using light from a source, relative to light from a reference source such as daylight or a blackbody of the same color temperature. By definition, an incandescent lamp has a CRI of 100. Real-life fluorescent tubes achieve CRIs of anywhere from 50 to 98. Fluorescent lamps with low CRI have phosphors that emit too little red light. Skin appears less pink, and hence "unhealthy" compared with incandescent lighting. Colored objects appear muted. For example, a low CRI 6800 K halophosphate tube (an extreme example) will make reds appear dull red or even brown. Since the eye is relatively less efficient at detecting red light, an improvement in color rendering index, with increased energy in the red part of the spectrum, may reduce the overall luminous efficacy.

Lighting arrangements use fluorescent tubes in an assortment of tints of white. Sometimes this is because of the lack of appreciation for the difference or importance of differing tube types. Mixing tube types within fittings can improve the color reproduction of lower quality tubes.

Phosphor Composition

Some of the least pleasant light comes from tubes containing the older, halophosphate-type phosphors (chemical formula $Ca_5(PO_4)_3(F, Cl):Sb^{3+}, Mn^{2+}$). This phosphor mainly emits yellow and blue light, and relatively little green and red. In the absence of a reference, this mixture appears white to the eye, but the light has an incomplete spectrum. The CRI of such lamps is around 60.

Since the 1990s, higher quality fluorescent lamps use either a higher CRI halophosphate coating, or a *triphosphor* mixture, based on europium and terbium ions, that have emission bands more evenly distributed over the spectrum of visible light. High CRI halophosphate and triphosphor tubes give a more natural color reproduction to the human eye. The CRI of such lamps is typically 82–100.

Fluorescent lamp spectra	
Typical fluorescent lamp with rare earth phosphor	A typical "cool white" fluorescent lamp utilizing two rare earth doped phosphors, Tb^{3+}, $Ce^{3+}:LaPO_4$ for green and blue emission and $Eu:Y_2O_3$ for red. For an explanation of the origin of the individual peaks click on the image. Several of the spectral peaks are directly generated from the mercury arc. This is likely the most common type of fluorescent lamp in use today.
An older style halophosphate phosphor fluorescent lamp	Halophosphate phosphors in these lamps usually consist of trivalent antimony and divalent manganese doped calcium halophosphate ($Ca_5(PO_4)_3(Cl, F):Sb^{3+}, Mn^{2+}$). The color of the light output can be adjusted by altering the ratio of the blue emitting antimony dopant and orange emitting manganese dopant. The color rendering ability of these older style lamps is quite poor. Halophosphate phosphors were invented by A.H. McKeag et al. in 1942.

"Natural sunshine" fluorescent light	Spectrum from a 48" Philips F32T8 natural sunshine fluorescent light	Peaks with stars are mercury-lines.
Yellow fluorescent lights		The spectrum is nearly identical to a normal fluorescent lamp except for a near total lack of light below 500 nanometers. This effect can be achieved through either specialized phosphor use or more commonly by the use of a simple yellow light filter. These lamps are commonly used as lighting for photolithography work in cleanrooms and as "bug repellent" outdoor lighting (the efficacy of which is questionable).
Spectrum of a "blacklight" lamp		There is typically only one phosphor present in a blacklight lamp, usually consisting of europium-doped strontium fluoroborate, which is contained in an envelope of Wood's glass.

Applications

Fluorescent lamps come in many shapes and sizes. The compact fluorescent lamp (CFL) is becoming more popular. Many compact fluorescent lamps integrate the auxiliary electronics into the base of the lamp, allowing them to fit into a regular light bulb socket.

In US residences, fluorescent lamps are mostly found in kitchens, basements, or garages, but schools and businesses find the cost savings of fluorescent lamps to be significant and rarely use incandescent lights. Tax incentives and building codes result in higher use in places such as California.

In other countries, residential use of fluorescent lighting varies depending on the price of energy, financial and environmental concerns of the local population, and acceptability of the light output. In East and Southeast Asia it is very rare to see incandescent bulbs in buildings anywhere.

Some countries are encouraging the phase-out of incandescent light bulbs and substitution of incandescent lamps with fluorescent lamps or other types of energy-efficient lamps.

In addition to general lighting, special fluorescent lights are often used in stage lighting for film and video production. They are cooler than traditional halogen light sources, and use high-frequency ballasts to prevent video flickering and high color-rendition index lamps to approximate daylight color temperatures.

Advantages

Luminous Efficacy

Fluorescent lamps convert more of the input power to visible light than incandescent lamps, though as of 2013 LEDs are sometimes even more efficient and are more rapidly increasing in efficiency. A typical 100 watt tungsten filament incandescent lamp may convert only 5% of its power input to visible white light (400–700 nm wavelength), whereas typical fluorescent lamps convert about 22% of the power input to visible white light.

The efficacy of fluorescent tubes ranges from about 16 lumens per watt for a 4 watt tube with an ordinary ballast to over 100 lumens per watt with a modern electronic ballast, commonly averaging 50 to 67 lm/W overall. Most compact fluorescents above 13 watts with integral electronic ballasts achieve about 60 lm/W. Lamps are rated by lumens after 100 hours of operation. For a given fluorescent tube, a high-frequency electronic ballast gives about a 10% efficacy improvement over an inductive ballast. It is necessary to include the ballast loss when evaluating the efficacy of a fluorescent lamp system; this can be about 25% of the lamp power with magnetic ballasts, and around 10% with electronic ballasts.

Fluorescent lamp efficacy is dependent on lamp temperature at the coldest part of the lamp. In T8 lamps this is in the center of the tube. In T5 lamps this is at the end of the tube with the text stamped on it. The ideal temperature for a T8 lamp is 25 °C (77 °F) while the T5 lamp is ideally at 35 °C (95 °F).

Life

Typically a fluorescent lamp will last 10 to 20 times as long as an equivalent incandescent lamp when operated several hours at a time. Under standard test conditions general lighting lamps have 9,000 hours or longer service life.

The higher initial cost of a fluorescent lamp compared with an incandescent lamp is usually more than compensated for by lower energy consumption over its life.

A few manufacturers are producing T8 lamps with 90,000 hour lamp lives, rivalling the life of LED lamps.

Lower Luminance

Compared with an incandescent lamp, a fluorescent tube is a more diffuse and physically larger light source. In suitably designed lamps, light can be more evenly distributed without point source of glare such as seen from an undiffused incandescent filament; the lamp is large compared to the typical distance between lamp and illuminated surfaces.

Lower Heat

Fluorescent lamps give off about one-fifth the heat of equivalent incandescent lamps. This greatly reduces the size, cost, and energy consumption devoted to air conditioning for office buildings that would typically have many lights and few windows.

Disadvantages

Frequent Switching

If the lamp is installed where it is frequently switched on and off, it will age rapidly. Under extreme conditions, its lifespan may be much shorter than a cheap incandescent lamp. Each start cycle slightly erodes the electron-emitting surface of the cathodes; when all the emission material is gone, the lamp cannot start with the available ballast voltage. Fixtures intended for flashing of lights (such as for advertising) will use a ballast that maintains cathode temperature when the arc is off, preserving the life of the lamp.

The extra energy used to start a fluorescent lamp is equivalent to a few seconds of normal operation; it is more energy-efficient to switch off lamps when not required for several minutes.

Health and Safety Issues

If a fluorescent lamp is broken, a very small amount of mercury can contaminate the surrounding environment. About 99% of the mercury is typically contained in the phosphor, especially on lamps that are near the end of their life. The broken glass is usually considered a greater hazard than the small amount of spilled mercury. The EPA recommends airing out the location of a fluorescent tube break and using wet paper towels to help pick up the broken glass and fine particles. Any glass and used towels should be disposed of in a sealed plastic bag. Vacuum cleaners can cause the particles to become airborne, and should not be used.

Fluorescent lamps with magnetic ballasts flicker at a normally unnoticeable frequency of 100 or 120 Hz and this flickering can cause problems for some individuals with light sensitivity; they are listed as problematic for some individuals with autism, epilepsy, lupus, chronic fatigue syndrome, Lyme disease, and vertigo. Newer fluorescent lights without magnetic ballasts have essentially eliminated flicker.

Ultraviolet Emission

Fluorescent lamps emit a small amount of ultraviolet (UV) light. A 1993 study in the US found that ultraviolet exposure from sitting under fluorescent lights for eight hours is equivalent to one minute of sun exposure. Ultraviolet radiation from compact fluorescent lamps may exacerbate symptoms in photosensitive individuals.

The ultraviolet light from a fluorescent lamp can degrade the pigments in paintings (especially watercolor pigments) and bleach the dyes used in textiles and some printing. Valuable art work must be protected from ultraviolet light by placing additional glass or transparent acrylic sheets between the lamp and the art work.

Ballast

Fluorescent lamps require a ballast to stabilize the current through the lamp, and to provide the initial striking voltage required to start the arc discharge. This increases the cost of fluorescent light fixtures, though often one ballast is shared between two or more lamps. Electromagnetic ballasts with a minor fault can produce an audible humming or buzzing noise. Magnetic ballasts are usually filled with a tar-like potting compound to reduce emitted noise. Hum is eliminated in lamps with a high-frequency electronic ballast. Energy lost in magnetic ballasts was around 10% of lamp input power according to GE literature from 1978. Electronic ballasts reduce this loss.

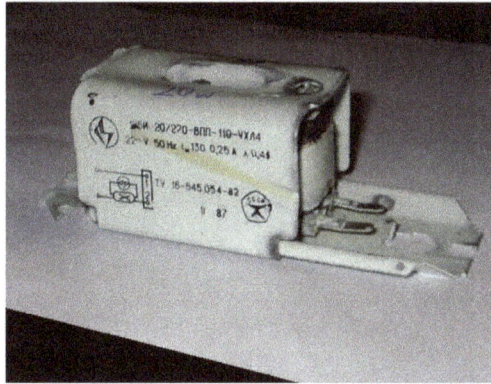

Magnetic single-lamp ballasts have a low power factor.

Power Quality and Radio Interference

Simple inductive fluorescent lamp ballasts have a power factor of less than unity. Inductive ballasts include power factor correction capacitors. Simple electronic ballasts may also have low power factor due to their rectifier input stage.

Fluorescent lamps are a non-linear load and generate harmonic currents in the electrical power supply. The arc within the lamp may generate radio frequency noise, which can be conducted through power wiring. Suppression of radio interference is possible. Very good suppression is possible, but adds to the cost of the fluorescent fixtures.

Operating Temperature

Fluorescent lamps operate best around room temperature. At much lower or higher temperatures, efficacy decreases. At below-freezing temperatures standard lamps may not start. Special lamps may be needed for reliable service outdoors in cold weather. In applications such as road and railway signalling, fluorescent lamps which do not generate as much heat as incandescent lamps may not melt snow and ice build up around the lamp, leading to reduced visibility.

Lamp Shape

Fluorescent tubes are long, low-luminance sources compared with high pressure arc lamps, incandescent lamps and LEDs. However, low luminous intensity of the emitting surface is useful because it reduces glare. Lamp fixture design must control light from a long tube instead of a compact globe.

The compact fluorescent lamp (CFL) replaces regular incandescent bulbs. However, some CFLs will not fit some lamps, because the harp (heavy wire shade support bracket) is shaped for the narrow neck of an incandescent lamp, while CFLs tend to have a wide housing for their electronic ballast close to the lamp's base.

Flicker Problems

Fluorescent lamps using a magnetic power line frequency ballast do not give out a steady light; instead, they flicker at twice the supply frequency. This results in fluctuations not only with light output but color temperature as well, which may pose problems for photography and people who are sensitive to the flicker. Even among persons not sensitive to light flicker, a stroboscopic effect can be noticed, where something spinning at just the right speed may appear stationary if illuminated solely by a single fluorescent lamp. This effect is eliminated by paired lamps operating on a lead-lag ballast. Unlike a true strobe lamp, the light level drops in appreciable time and so substantial "blurring" of the moving part would be evident.

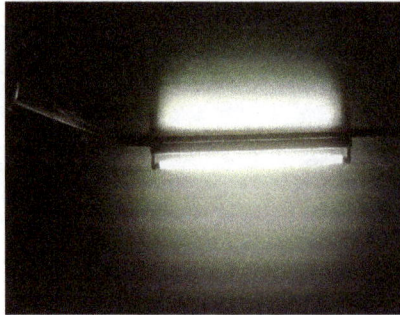

The "beat effect" problem created when shooting photos under standard fluorescent lighting

In some circumstances, fluorescent lamps operated at the power supply frequency (50 or 60 Hz) can also produce flicker at the same frequency itself, which is noticeable by more people. This can happen in the last few hours of tube life when the cathode emission coating at one end has almost run out, and that cathode starts having difficulty emitting enough electrons into the gas fill, resulting in slight rectification and hence uneven light output in positive and negative going AC cycles. Power frequency flicker can also sometimes be emitted from the very ends of the tubes, if each tube electrode produces a slightly different light output pattern on each half-cycle. Flicker at power frequency is more noticeable in the peripheral vision than it is when viewed directly, as is all flicker (since the peripheral vision is faster—has a higher critical frequency—than the central vision).

Near the end of life, fluorescent lamps can start flickering at a frequency lower than the power frequency. This is due to a dynamic instability inherent in the negative resistance of the plasma source, which can be from a bad lamp, a bad ballast, or a bad starter; or occasionally from a poor connection to power.

New fluorescent lamps may show a twisting spiral pattern of light in a part of the lamp. This effect is due to loose cathode material and usually disappears after a few hours of operation.

Electromagnetic ballasts may also cause problems for video recording as there can be a so-called *beat effect* between the periodic reading of a camera's sensor and the fluctuations in intensity of the fluorescent lamp.

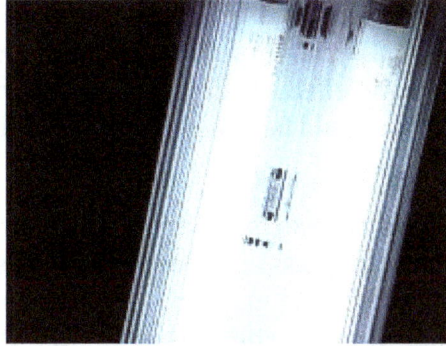

The "beat effect" problem created when shooting films under standard fluorescent lighting

Fluorescent lamps using high-frequency electronic ballasts do not produce visible light flicker, since above about 5 kHz, the excited electron state half-life is longer than a half cycle, and light production becomes continuous. Operating frequencies of electronic ballasts are selected to avoid interference with infrared remote controls. Poor quality (or failing) electronic ballasts may have insufficient reservoir capacitance or have poor regulation, thereby producing considerable 100/120 Hz modulation of the light.

Dimming

Fluorescent light fixtures cannot be connected to dimmer switches intended for incandescent lamps. Two effects are responsible for this: the waveform of the voltage emitted by a standard phase-control dimmer interacts badly with many ballasts, and it becomes difficult to sustain an arc in the fluorescent tube at low power levels. Dimming installations require a compatible dimming ballast. These systems keep the cathodes of the fluorescent tube fully heated even as the arc current is reduced, promoting easy thermionic emission of electrons into the arc stream. CFLs are available that work in conjunction with a suitable dimmer.

Disposal and Recycling

The disposal of phosphor and particularly the toxic mercury in the tubes is an environmental issue. Governmental regulations in many areas require special disposal of fluorescent lamps separate from general and household wastes. For large commercial or industrial users of fluorescent lights, recycling services are available in many nations, and may be required by regulation. In some areas, recycling is also available to consumers.

Lamp Sizes and Designations

Systematic nomenclature identifies mass-market lamps as to general shape, power rating, length, color, and other electrical and illuminating characteristics.

Other Fluorescent Lamps

Black lights

Blacklights are a subset of fluorescent lamps that are used to provide near ultraviolet light (at about 360 nm wavelength). They are built in the same fashion as conventional fluorescent lamps but the glass tube is coated with a phosphor that converts the short-wave UV within the tube to long-wave UV rather than to visible light. They are used to provoke fluorescence (to provide dramatic effects using blacklight paint and to detect materials such as urine and certain dyes that would be invisible in visible light) as well as to attract insects to bug zappers.

So-called *blacklite blue* lamps are also made from more expensive deep purple glass known as Wood's glass rather than clear glass. The deep purple glass filters out most of the visible colors of light directly emitted by the mercury-vapor discharge, producing proportionally less visible light compared with UV light. This allows UV-induced fluorescence to be seen more easily (thereby allowing blacklight posters to seem much more dramatic). The blacklight lamps used in bug zappers do not require this refinement so it is usually omitted in the interest of cost; they are called simply *blacklite* (and not blacklite blue).

Tanning lamps

The lamps used in tanning beds contain a different phosphor blend (typically 3 to 5 or more phosphors) that emits both UVA and UVB, provoking a tanning response in most human skin. Typically, the output is rated as 3–10% UVB (5% most typical) with the remaining UV as UVA. These are mainly F71, F72, or F73 HO (100 W) lamps, although 160 W VHO are somewhat common. One common phosphor used in these lamps is lead-activated barium disilicate, but a europium-activated strontium fluoroborate is also used. Early lamps used thallium as an activator, but emissions of thallium during manufacture were toxic.

UVB medical lamps

The lamps used in phototherapy contain a phosphor that emits only UVB ultraviolet light. There are two types: broadband UVB that gives 290–320 nanometer with peak wavelength of 306 nm, and narrowband UVB that gives 311–313 nanometer. Due to its longer wavelength the narrowband UVB requires a 10 times higher dose to the skin, compared to the broadband. The narrowband is good for psoriasis, eczema (atopic dermatitis), vitiligo, lichen planus, and some other skin diseases. The broadband is better for increasing Vitamin D3 in the body.

Grow lamps

Grow lamps contain phosphor blends that encourage photosynthesis, growth, or flowering in plants, algae, photosynthetic bacteria, and other light-dependent organisms. These often emit light primarily in the red and blue color range, which is absorbed by chlorophyll and used for photosynthesis in plants.

Infrared lamps

Lamps can be made with a lithium metaluminate phosphor activated with iron. This phosphor has peak emissions between 675 and 875 nanometers, with lesser emissions in the deep red part of the visible spectrum.

Bilirubin lamps

Deep blue light generated from a europium-activated phosphor is used in the light therapy treatment of jaundice; light of this color penetrates skin and helps in the breakup of excess bilirubin.

Germicidal lamps - (similar structure but NOT fluorescent)

Germicidal lamps depend on the property that spectrum of 254 nm kills most germs. Germicidal lamps contain no phosphor at all (making them mercury vapor gas discharge lamps rather than fluorescent) and their tubes are made of fused quartz that is transparent to the UV light emitted by the mercury discharge. The 254 nm UV emitted by these tubes will kill germs and ionize oxygen to ozone. In addition it can cause eye and skin damage and should not be used or observed without eye and skin protection. Besides their uses to kill germs and create ozone, they are sometimes used by geologists to identify certain species of minerals by the color of their fluorescence. When used in this fashion, they are fitted with filters in the same way as blacklight-blue lamps are; the filter passes the short-wave UV and blocks the visible light produced by the mercury discharge. They are also used in some EPROM erasers.

Germicidal lamps have designations beginning with G (meaning *germicidal*), rather than F, for example G30T8 for a 30-watt, 1-inch (2.5 cm) diameter, 36-inch (91 cm) long germicidal lamp (as opposed to an F30T8, which would be the fluorescent lamp of the same size and rating).

Electrodeless lamps

Electrodeless induction lamps are fluorescent lamps without internal electrodes. They have been commercially available since 1990. A current is induced into the gas column using electromagnetic induction. Because the electrodes are usually the life-limiting element of fluorescent lamps, such electrodeless lamps can have a very long service life, although they also have a higher purchase price.

Cold-cathode fluorescent lamps (CCFL)

Cold-cathode fluorescent lamps are used as backlighting for LCDs in personal computer and TV monitors. They are also popular with computer case modders in recent years.

Science Demonstrations

Capacitive coupling with high-voltage power lines can light a lamp continuously at low intensity.

Fluorescent lamps can be illuminated by means other than a proper electrical connection. These other methods, however, result in very dim or very short-lived illumination, and so are seen mostly in science demonstrations. Static electricity or a Van de Graaff generator will cause a lamp to flash momentarily as it discharges a high voltage capacitance. A Tesla coil will pass high-frequency current through the tube, and since it has a high voltage as well, the gases within the tube will ionize and emit light. Capacitive coupling with high-voltage power lines can light a lamp continuously at low intensity, depending on the intensity of the electrostatic field, as shown in the image on the right.

Battery Charger

This unit charges the batteries until they reach a specific voltage and then it trickle charges the batteries until it is disconnected.

A simple charger equivalent to an AC/DC wall adapter. It applies 300mA to the battery at all times, which will damage the battery if left connected too long.

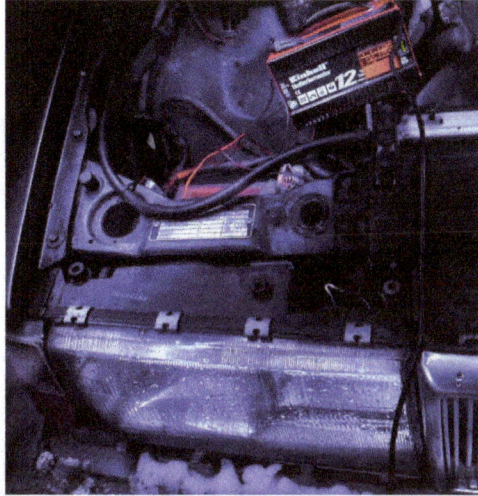
Car Battery Charger

A battery charger, or recharger, is a device used to put energy into a secondary cell or rechargeable battery by forcing an electric current through it.

The charging protocol depends on the size and type of the battery being charged. Some battery types have high tolerance for overcharging and can be recharged by connection to a constant voltage source or a constant current source; simple chargers of this type require manual disconnection at the end of the charge cycle, or may have a timer to cut off charging current at a fixed time. Other battery types cannot withstand long high-rate over-charging; the charger may have temperature or voltage sensing circuits and a microprocessor controller to adjust the charging current, determine the state of charge, and cut off at the end of charge.

A trickle charger provides a relatively small amount of current, only enough to counteract self-discharge of a battery that is idle for a long time. Slow battery chargers may take several hours to complete a charge; high-rate chargers may restore most capacity within minutes or less than an hour, but generally require monitoring of the battery to protect it from overcharge. Electric vehicles need high-rate chargers for public access; installation of such chargers and the distribution support for them is an issue in the proposed adoption of electric cars.

C-Rates

Charge and discharge rates are often denoted as C or *C-rate*, which is a measure of the rate at which a battery is charged or discharged relative to the capacity of the battery. The C-rate is given by the numerical value of the ratio of the charging or discharging current in A to the capacity of the battery in Ah.

For example, for a battery with a capacity of 500mAh, the current corresponding to a C-rate of 10 and a charge or discharge time of 6 minutes would be 5000mA or 5A, while the current corresponding to a C-rate of 1/2 and a charge or discharge time of 2 hours would be 250mA.

Very rapid charging rates, 1 hour or less, generally require the charger to carefully monitor battery parameters such as terminal voltage and temperature to prevent overcharging and damage to the cells.

Types of Battery Chargers

Simple Chargers

A typical simple charger

A simple charger works by supplying a constant DC or pulsed DC power source to a battery being charged. The simple charger does not alter its output based on time or the charge on the battery. This simplicity means that a simple charger is inexpensive, but there is a tradeoff in quality. Typically, a simple charger takes longer to charge a battery to prevent severe over-charging. Even so, a battery left in a simple charger for too long will be weakened or destroyed due to over-charging. These chargers can supply either a constant voltage or a constant current to the battery.

Simple AC-powered battery chargers have much higher ripple current and ripple voltage than other kinds of battery supplies. When the ripple current is within the battery-manufacturer-recommended level, the ripple voltage will also be well within the recommended level. The maximum ripple current for a typical 12 V 100 Ah VRLA battery is 5 amps. As long as the ripple current is not excessive (more than 3 to 4 times the battery-manufacturer-recommended level), the expected life of a ripple-charged VRLA battery is within 3% of the life of a constant DC-charged battery.

Fast Chargers

Fast chargers make use of control circuitry in the batteries being charged to rapidly charge the batteries without damaging the cells' elements. Most such chargers have a cooling fan to help keep the temperature of the cells under control. Most are also capable of acting as standard overnight chargers if used with standard NiMH cells that do not have the special control circuitry.

Inductive Chargers

Inductive battery chargers use electromagnetic induction to charge batteries. A charging station sends electromagnetic energy through inductive coupling to an electrical device, which stores the energy in the batteries. This is achieved without the need for metal contacts between the charger and the battery. It is commonly used in electric toothbrushes and other devices used in bathrooms. Because there are no open electrical contacts, there is no risk of electrocution.

Intelligent Chargers

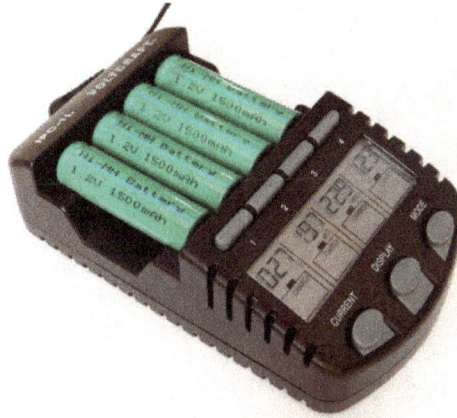

Example of a smart charger for AA and AAA batteries

A "smart charger" should not be confused with a "smart battery". A smart battery is generally defined as one containing some sort of electronic device or "chip" that can communicate with a smart charger about battery characteristics and condition. A smart battery generally requires a smart charger it can communicate with. A smart charger is defined as a charger that can respond to the condition of a battery, and modify its charging actions accordingly.

Some smart chargers are designed to charge:

- "smart" batteries.

- "dumb" batteries, which lack any internal electronic circuitry.

The output current of a smart charger depends upon the battery's state. An intelligent charger may monitor the battery's voltage, temperature or time under charge to determine the optimum charge current and to terminate charging.

For Ni-Cd and NiMH batteries, the voltage across the battery increases slowly during the charging process, until the battery is fully charged. After that, the voltage *decreases*, which indicates to an intelligent charger that the battery is fully charged. Such chargers are often labeled as a ΔV, "delta-V," or sometimes "delta peak", charger, indicating that they monitor the voltage change.

The problem is, the magnitude of "delta-V" can become very small or even non-existent if (very) high capacity rechargeable batteries are recharged. This can cause even an intelligent battery charger to not sense that the batteries are actually already fully charged, and continue charging. Overcharging of the batteries will result in some cases. However, many so called intelligent chargers employ a combination of cut off systems, which should prevent overcharging in the vast majority of cases.

A typical intelligent charger fast-charges a battery up to about 85% of its maximum capacity in less than an hour, then switches to trickle charging, which takes several hours to top off the battery to its full capacity.

Motion-Powered Charger

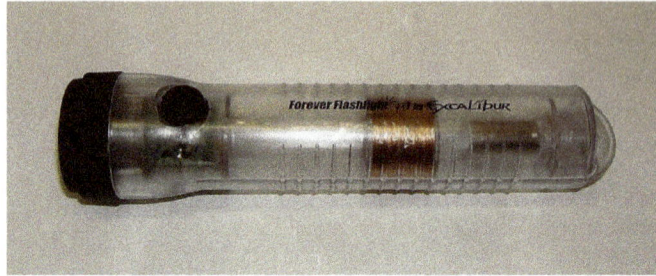

Linear induction flashlight, charged by shaking along its long axis, causing magnet *(visible at right)* to slide through a coil of wire *(center)* to generate electricity

Several companies have begun making devices that charge batteries based on regular human motion. One example, made by Tremont Electric, consists of a magnet held between two springs that can charge a battery as the device is moved up and down, such as when walking. Such products have not yet achieved significant commercial success.

Pulse Chargers

Some chargers use *pulse technology* in which a series of voltage or current pulses is fed to the battery. The DC pulses have a strictly controlled rise time, pulse width, pulse repetition rate (frequency) and amplitude. This technology is said to work with any size, voltage, capacity or chemistry of batteries, including automotive and valve-regulated batteries. With pulse charging, high instantaneous voltages can be applied without overheating the battery. In a Lead–acid battery, this breaks down lead-sulfate crystals, thus greatly extending the battery service life.

Several kinds of pulse charging are patented. Others are open source hardware.

Some chargers use pulses to check the current battery state when the charger is first connected, then use constant current charging during fast charging, then use pulse charging as a kind of trickle charging to maintain the charge.

Some chargers use "negative pulse charging", also called "reflex charging" or "burp charging". Such chargers use both positive and brief negative current pulses. There is no significant evidence, however, that negative pulse charging is more effective than ordinary pulse charging.

Solar Chargers

Solar chargers convert light energy into DC current. They are generally portable, but can also be fixed mount. Fixed mount solar chargers are also known as solar panels. Solar panels are often connected to the electrical grid, whereas portable solar chargers are used off-the-grid (i.e. cars, boats, or RVs).

Timer-Based(HI) Chargers

The output of a timer charger is terminated after a pre-determined time. Timer chargers were the most common type for high-capacity Ni-Cd cells in the late 1990s for example (low-capacity consumer Ni-Cd cells were typically charged with a simple charger).

Often a timer charger and set of batteries could be bought as a bundle and the charger time was set to suit those batteries. If batteries of lower capacity were charged then they would be overcharged, and if batteries of higher capacity were charged they would be only partly charged. With the trend for battery technology to increase capacity year on year, an old timer charger would only partly charge the newer batteries.

Timer based chargers also had the drawback that charging batteries that were not fully discharged, even if those batteries were of the correct capacity for the particular timed charger, would result in over-charging.

Trickle Chargers

A trickle charger is typically a low-current (5–1,500 mA) battery charger. A trickle charger is generally used to charge small capacity batteries (2–30 Ah). These types of battery chargers are also used to maintain larger capacity batteries (> 30 Ah) that are typically found on cars, boats, RVs and other related vehicles. In larger applications, the current of the battery charger is sufficient only to provide a maintenance or trickle current (trickle is commonly the last charging stage of most battery chargers). Depending on the technology of the trickle charger, it can be left connected to the battery indefinitely. Some battery chargers that can be left connected to the battery without causing the battery damage are also referred to as smart or intelligent chargers.

Universal Battery Charger–Analyzers

The most sophisticated types are used in critical applications (e.g. military or aviation batteries). These heavy-duty automatic "intelligent charging" systems can be programmed with complex charging cycles specified by the battery maker. The best are universal (i.e. can charge all battery types), and include automatic capacity testing and analyzing functions too.

USB-Based Chargers

Australian and New Zealand power socket with USB charger socket

Since the Universal Serial Bus specification provides for a five-volt power supply, it is possible to use a USB cable to connect a device to a power supply. Products based on this approach include chargers for cellular phones, portable digital audio players, and tablet computers. They may be fully compliant USB peripheral devices adhering to USB power discipline, or uncontrolled in the manner of USB decorations.

Although portable solar chargers obtain energy from the sun only, they still can (depending on the technology) be used in low light (i.e. cloudy) applications. Portable solar chargers are typically used for trickle charging, although some solar chargers (depending on the wattage), can completely recharge batteries. Other devices may exist, which combine this with other sources of energy for added recharging efficacy.

Powerbank

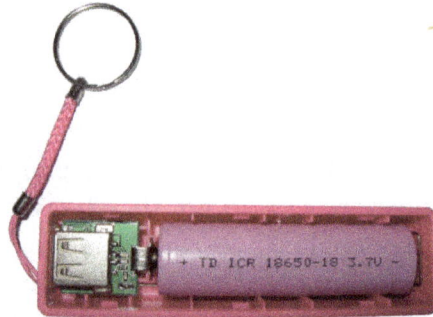

A typical USB powerbank with the cover removed. The internal 18650 size lithium-ion battery is exposed.

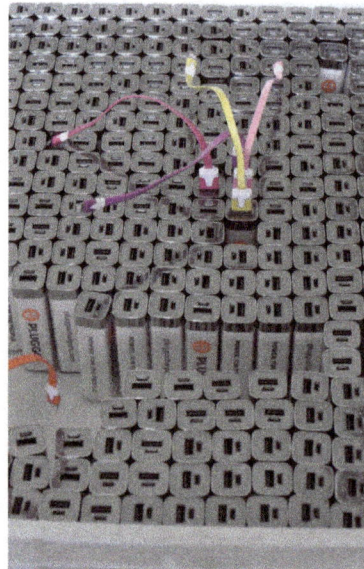

Powerbanks with cover

Powerbanks are popular for charging smartphones and mobile tablet devices. A powerbank is a portable device that can supply power from its built-in batteries through a USB port. They usually recharge with USB power supply. Technically, a powerbank consists of rechargeable Lithium-ion or Lithium-Polymer batteries installed in a protective casing, guided by a printed circuit board (PCB) which ensures various protective and safety measures.

Specifications:

- Capacity in mAh: mAh stands for milli Ampere-hour and measures the amount of power flow that can be supplied by a certain powerbank. Amount of mA × time at 5V ideally. Many manufacturers measure this at the voltage of battery inside, hence they show more than actual.

- Simultaneous charging and discharging: need to specify if the powerbank can be used while it is charging.

- Number of output USB ports: This specifies the number of devices that can be charged simultaneously.

- Output current rating: This specifies the current rating that it can charge maximum. The higher the number, the better the powerbank. This can vary from output port to output port.

- Input Current Rating: Input current rating is the amount of current the powerbank is able to draw at its maximum level while getting charged.

- Safety Protections: Over Voltage Protection, Over Charge Protections, Over Current Protections, Over Heat Protections, Short-Circuit Protections and Over Discharge Protections are the common safety measures observed with standard powerbanks.

- LED Indications: The Led glows as per indicating the amount of charging ability left with the powerbank.

Applications

Since a battery charger is intended to be connected to a battery, it may not have voltage regulation or filtering of the DC voltage output. Battery chargers equipped with both voltage regulation and filtering may be identified as battery eliminators.

Battery Charger for Vehicles

There are two main types of charges for vehicles:

- To recharge a fuel vehicle's starter battery, where a modular charger is used; typically an 3-stage charger.

- To recharge an electric vehicle (EV) battery pack.

Chargers for car batteries come in varying ratings. Chargers that are rated up to two amperes may be used to maintain charge on parked vehicle batteries or for small batteries on garden tractors or similar equipment. A motorist may keep a charger rated a few amperes to ten or fifteen amperes for maintenance of automobile batteries or to recharge a vehicle battery that has accidentally discharged. Service stations and commercial garages will have a large charger to fully charge a battery in an hour or two; often these chargers can briefly source the hundreds of amperes required to crank an internal combustion engine starter.

Electric Vehicle Batteries

Electric vehicle battery chargers come in a variety of brands and characteristics. Zivan, Manzanita Micro, Elcon, Quick Charge, Rossco, Brusa, Delta-Q, Kelly, Lester and Soneil are the top 10

EV chargers in 2011 according to EVAlbum.com. These chargers vary from 1 kW to 7.5 kW maximum charge rate. Some use algorithm charge curves, others use constant voltage, constant current. Some are programmable by the end user through a CAN port, some have dials for maximum voltage and amperage, some are preset to specified battery pack voltage, amp-hour and chemistry. Prices range from $400 to $4500.

A 10 amp-hour battery could take 15 hours to reach a fully charged state from a fully discharged condition with a 1 amp charger as it would require roughly 1.5 times the battery's capacity.

Public EV charging stations provide 6 kW (host power of 208 to 240 VAC off a 40 amp circuit). 6 kW will recharge an EV roughly 6 times faster than 1 kW overnight charging.

Rapid charging results in even faster recharge times and is limited only by available AC power and the type of charging system.

Onboard EV chargers (change AC power to DC power to recharge the EV's pack) can be:

- Isolated: they make no physical connection between the A/C electrical mains and the batteries being charged. These typically employ some form of Inductive charging. Some isolated chargers may be used in parallel. This allows for an increased charge current and reduced charging times. The battery has a maximum current rating that cannot be exceeded

- Non-isolated: the battery charger has a direct electrical connection to the A/C outlet's wiring. Non-isolated chargers cannot be used in parallel.

Power Factor Correction (PFC) chargers can more closely approach the maximum current the plug can deliver, shortening charging time.

Charge Stations

Project Better Place was deploying a network of charging stations and subsidizing vehicle battery costs through leases and credits until filing for bankruptcy in May 2013.

Auxiliary charger designed to fit a variety of proprietary devices

Non-Contact Magnetic Charging

Researchers at the Korea Advanced Institute of Science and Technology (KAIST) have developed an electric transport system (called Online Electric Vehicle, OLEV) where the vehicles get their power needs from cables underneath the surface of the road via non-contact magnetic charging, (where a power source is placed underneath the road surface and power is wirelessly picked up on the vehicle itself.

Mobile Phone Charger

Micro USB mobile phone charger

Pay-per-charge kiosk, illustrating the variety of mobile phone charger connectors

Most mobile phone chargers are not really chargers, only power adapters that provide a power source for the charging circuitry which is almost always contained within the mobile phone. Older ones are notoriously diverse, having a wide variety of DC connector-styles and voltages, most of which are not compatible with other manufacturers' phones or even different models of phones from a single manufacturer.

Users of publicly accessible charging kiosks must be able to cross-reference connectors with device brands/models and individual charge parameters and thus ensure delivery of the correct charge for their mobile device. A database-driven system is one solution, and is being incorporated into some designs of charging kiosks.

Mobile phones can usually accept a relatively wide range of voltages, as long as it is sufficiently above the phone battery's voltage. However, if the voltage is too high, it can damage the phone. Mostly, the voltage is 5 volts or slightly higher, but it can sometimes vary up to 12 volts when the power source is not loaded..

There are also human-powered chargers sold on the market, which typically consists of a dynamo powered by a hand crank and extension cords. A French startup offers a kind of dynamo charger inspired by the ratchet that can be used with only one hand. There are also solar chargers, including one that is a fully mobile personal charger and panel, which you can easily transport.

China, the European Commission and other countries are making a national standard on mobile phone chargers using the USB standard. In June 2009, 10 of the world's largest mobile phone manufacturers signed a Memorandum of Understanding to develop specifications for and support a microUSB-equipped common External Power Supply (EPS) for all data-enabled mobile phones sold in the EU. On October 22, 2009, the International Telecommunication Union announced a standard for a universal charger for mobile handsets (Micro-USB).

Stationary Battery Plants

Telecommunications, electric power, and computer uninterruptible power supply facilities may have very large standby battery banks (installed in battery rooms) to maintain critical loads for several hours during interruptions of primary grid power. Such chargers are permanently installed and equipped with temperature compensation, supervisory alarms for various system faults, and often redundant independent power supplies and redundant rectifier systems. Chargers for stationary battery plants may have adequate voltage regulation and filtration and sufficient current capacity to allow the battery to be disconnected for maintenance, while the charger supplies the DC system load. Capacity of the charger is specified to maintain the system load and recharge a completely discharged battery within, say, 8 hours or other interval.

Use in Experiments

A battery charger can work as a DC power adapter for experimentation. It may, however, require an external capacitor to be connected across its output terminals in order to "smooth" the voltage sufficiently, which may be thought of as a DC voltage plus a "ripple" voltage added to it. There may be an internal resistance connected to limit the short circuit current, and the value of that internal resistance may have to be taken into consideration in experiments.

Prolonging Battery Life

What practices are best depend on the type of battery. NiCd cells need to be fully discharged occasionally, or else the battery loses capacity over time in a phenomenon known as "memory effect." Once a month (once every 30 charges) is sometimes recommended. This extends the life of the battery since memory effect is prevented while avoiding full charge cycles which are known to be hard on all types of dry-cell batteries, eventually resulting in a permanent decrease in battery capacity.

Most modern cell phones, laptops, and most electric vehicles use Lithium-ion batteries. These batteries last longest if the battery is frequently charged; fully discharging them will degrade their capacity relatively quickly. When storing however, lithium batteries degrade more while fully charged than if they are only 40% charged. Degradation also occurs faster at higher temperatures. Degradation in lithium-ion batteries is caused by an increased internal battery resistance due to cell oxidation. This decreases the efficiency of the battery, resulting in less net current available to be drawn from the battery. However, if Li-ION cells are discharged below a certain voltage a chemical reaction occurs that make them dangerous if recharged, which is why probably all such batteries in consumer goods now have an "electronic fuse" that permanently disables them if the voltage falls below a set level. The electronic fuse draws a small amount of current from the battery, which means that if a laptop battery is left for a long time without charging it, and with a very low initial state of charge, the battery may be permanently destroyed.

Motor vehicles, such as boats, RVs, ATVs, motorcycles, cars, trucks, and more use lead–acid batteries. These batteries employ a sulfuric acid electrolyte and can generally be charged and discharged without exhibiting memory effect, though sulfation (a chemical reaction in the battery which deposits a layer of sulfates on the lead) will occur over time. Typically sulfated batteries are simply replaced with new batteries, and the old ones recycled. Lead–acid batteries will experience substantially longer life when a maintenance charger is used to "float charge" the battery. This prevents the battery from ever being below 100% charge, preventing sulfate from forming. Proper temperature compensated float voltage should be used to achieve the best results.

References

- Gribben, John; "The Scientists; A History of Science Told Through the Lives of Its Greatest Inventors"; Random House; 2004; pp 424–432; ISBN 978-0-8129-6788-3

- Weeks, Mary Elvira (2003). Discovery of the Elements: Third Edition (reprint). Kessinger Publishing. p. 287. ISBN 978-0-7661-3872-8.

- van Dulken, Stephen (2002). Inventing the 20th century: 100 inventions that shaped the world : from the airplane to the zipper. New York University Press. p. 42. ISBN 978-0-8147-8812-7.

- Kane, Raymond; Sell, Heinz (2001). "A Review of Early Inorganic Phosphors". Revolution in lamps: a chronicle of 50 years of progress. p. 98. ISBN 978-0-88173-378-5.

- William M. Yen, Shigeo Shionoya, Hajime Yamamoto, Practical Applications of Phosphors,CRC Press, 2006, ISBN 1-4200-4369-2, pages 84-85

- Van Broekhoven, Jacob (2001). "Chapter 5: Lamp Phosphors". In Kane, Raymond; Sell, Heinz. Revolution in lamps: a chronicle of 50 years of progress (2nd ed.). The Fairmont Press, Inc. p. 93. ISBN 0-88173-378-4.

- Fink, Donald G.; Beaty, H. Wayne, eds. (1978). Standard Handbook for Electrical Engineers (11th ed.). McGraw Hill. pp. 22–17. ISBN 978-0-070-20974-9.

- SCENIHR (Scientific Committee on Emerging and Newly-Identified Health Risks) (23 September 2008). "Scientific opinion on light sensitivity" (PDF). Retrieved 16 January 2016.

- "Compact Fluorescent Lighting" (PDF). eere.energy.gov. Archived from the original (PDF) on May 11, 2011. Retrieved 24 July 2012.

- "Science Fact or Science Fiction: Fluorescent Lights". Quirks and Quarks. CBC. Archived from the original on October 28, 2011. Retrieved 27 October 2011.

Permissions

Index

www.ingramcontent.com/pod-product-compliance
Lightning Source LLC
Chambersburg PA
CBHW061314190326
41458CB00011B/3802